LET'S RESTORE OUR LAND

Written by:

Team for Care of the Creation

Edited by:

Dan Fountain

Published by:

ECHO
North Fort Myers, Florida
www.echonet.org

REPRODUCTIONS AND PERMISSIONS

Dan Fountain, M.D.
1805 Turban Court
Fort Myers, FL 33908, USA
defountain@king.edu

Published by:
ECHO, 17391 Durrance Road, North Fort Myers, FL 33917-2212
ISBN 978-1-946263-28-5

TABLE OF CONTENTS

Dear Brothers and Sisters,

God created a wonderful world. He covered the earth with forest, grassland, and water. He made many kinds of animals, birds, and fish. Finally he made a man and a woman and told them to rule over the earth with Him. The forest, land, and water provided sufficient food and resources for all that God had made.

Soon, however, the man and woman decided to turn away from God and do what they wanted to do with the land, the forest, and the animals. After many years the forest began to disappear. The land no longer produced enough food. Many springs of water and even streams and lakes dried up. Today, many of the marvelous resources God created for us are gone. We often go hungry and have great difficulty in getting enough food and water to live.

Although we turned away from God, he did not tum away from us. He gave us a book that provides much wisdom for us about how to live, even about how to farm. We call it the Bible. God actually became a man called Jesus who lived among us, taught us how to work together in peace, and made it possible for us to return to our close relationship with God. God gave us good minds to think, to study nature, and to discover, by science, the laws God put into all of nature.

This book is about a pastor and his wife who saw what was happening to the land. They believed God was not pleased at what we are doing to the land. They decided to find out from God and from what they could learn by observing the land how we are supposed to care for the land. They began making changes, and then helping others in the church make changes. Soon they had more food to eat, good water to drink, and better health. People from surrounding communities became interested, and asked the pastor for help. By working together, learning from God, from science, and from each other, they began to restore the land.

Look at your community and the land around it. What do you see happening to the trees and forest? Is your land producing much food? Are animals providing food and resources for you? Or do you see that there are problems with all of this? If so this book is for you and for your community.

This is not a book just to read. It is a book to study, and to study with your family, friends, and neighbors. Read what Pastor Simon says. See what he has done and listen to what he discusses with others. Then study your own situation. Try to determine how Pastor Simon set out to solve problems of the land, the forest, and springs of water. Learn from him how to solve similar problems you may have. Begin discussing this with others to discover what you can do together to restore the land. As we restore the land, we begin to find the better life that God wants us to have.

Pastor Simon found in the Bible important principles about life and farming. He even found practical suggestions about the land, water, and animals. He likewise observed the land and discovered ways to make it fertile again. Pastor Simon's nephew John, who is studying agriculture, also helped him find better ways to care for the land.

At the end of each chapter are questions for study. Think about these questions yourself and discuss them with others. They can help you find answers as to how best to improve the land and resources God has given us. Then talk about what you all can do together to care for the land, the forest, and indeed, all of creation.

The people in this story are:

- Simon, pastor of the church in Katindi
- His wife Virginia
- John, his nephew who is studying in an agricultural school
- Moses, headman of the community at Malenga
- Robert, director of the secondary school in Katindi
- Rebecca, a school teacher
- Yvonne, a nurse
- Roger, a carpenter in Katindi
- David, a farmer from Katindi
- Many others from Katindi and the communities around it

Read about what they learned from Pastor Simon and each other. See how they studied the problems and what could be done to resolve the problems. Look at the different methods they used to improve their farming, control the animals, restore the forest, and improve the economy in their community. They learned from God, the Bible he has given us, and from the discoveries of science how to improve our lives and restore our land.

Let us follow their example before the destruction of the land and forest has gone further. If that continues, it will become impossible to restore them. So now is the time to restore the land and its many resources that God has given us. By working together and getting wisdom from God and science, we can succeed in restoring the land and improving our lives.

Many people have worked together to write this book. Some of them live and work in Africa. One farmer in Zimbabwe teaches farmers there about how to farm God's way. Some of his ideas are in this book. You can learn about him on the website www.farming-gods-way.org

Another group works in East Africa. You can learn of their work from the website www.careofcreation.org

Still another group works in the Democratic Republic of the Congo. Their website is www.sanru.org

A group of experts in agriculture in the United States of America works at an agricultural research and training center called ECHO. Many of the ideas from them are in this book. Their website is www.echonet.org

If you have access to the internet, you can find many good things from these places. All of them are working together to teach God's wisdom to people who want to make their lives better, to have plenty of good food, and to know the life that God gives us through his Son Jesus Christ who wants us all to have an abundant life.

Sincerely

The church development committee

Chapter 1

The Land is God's, but Ours to Use and Protect

The town of Katindi is on a low hill. About one thousand people live in the town. Katindi is surrounded by the fields and gardens of the people of the community. Further away are hills, some of them quite high. Many years ago forest covered those hills. Much of that forest is now gone, replaced by small farms and fields where cattle graze. Deep gullies from soil erosion scar the hills in many places.

In the town of Katindi is a large church. The pastor of the church, Simon Mavula, works hard to help his people understand the Word of God. He tells people that God's Word is very important for every aspect of our lives. Many members of his congregation meet with him often to talk about how they can make their lives better. They realize that God wants us to be healthy, to have plenty of food, and to have the abundant life that Jesus came to give us (see John 10:10). They also understand that, for these good things to happen, they must work hard. They must understand God's Word and his laws and put them into practice.

A few years ago, a group of church members met one evening at the home of Pastor Simon for a Bible study. They were discussing the study of the Garden of Eden that we find in the second chapter of Genesis, the first book of the Bible. David, a farmer, spoke up.

"Pastor, it says that God planted a garden in Eden. Was God a farmer?"

Pastor Simon and members of his church.

Was God a farmer?

"He is a farmer, indeed," said Pastor Simon. "Who do you think planted all the trees in the world before he created man? Did maize grow then, and millet, and rice, and wheat? The Bible doesn't talk about that, but I am sure those things existed before he made man. God created those foods and planted them here on the earth for us. So you see, David, if God is a farmer, then farming is a wonderful occupation."

"Wow! I never thought about it that way," David said.

Rebecca, a teacher, said, "It says that God planted fruit trees and other kinds of trees for food. Is that right?"

"Of course it's right," the Pastor said. 'The Bible always tells us the truth. In fact, God made hundreds of kinds of fruit trees, many of which grow right here where we are. We could have many more kinds of fruit if we knew how to grow them."

Robert, the school director, said, "In addition to that, God made many thousands of other kinds of trees–trees for the forest, others for grasslands, some for hills and mountains, others for valleys. He made a tremendous variety of trees. They all grow well together to make the earth beautiful and productive."

Arthur objected, "I sure don't see many trees around here. Where are all the fruit trees and other kinds of trees God made?"

Pastor Simon read verse 15 of Genesis chapter 2 to them: 'The Lord God took the man and put him in the Garden of Eden to work it and to take care of it'

"The problem is this," he said. "We have not taken very good care of the earth, the land, the trees, the forest, and the foods God has made. That is why we do not have the abundant life Jesus wants us to have. This is something very serious for us to think about I will be speaking about this in the morning church service on Sunday."

Sunday morning many people came to church. They wanted to hear what the pastor was going to say about the forest and the land.

He read a parable Jesus told about a man who planted a vineyard. The man then rented it to other people to take care of it. We read this parable in the Gospel of Mark, chapter 12. It goes like this:

> *Once there was a man who planted a vineyard, put a wall around it, dug a hole for the wine press, and built a watchtower. Then he rented the vineyard to tenants and left home on a trip. When the time came to gather the grapes, he sent a slave to the tenants to receive from them his share*

of the harvest. The tenants grabbed the slave, beat him, and sent him back without a thing. Then the owner sent another slave; the tenants beat him over the head and treated him shamefully. The owner sent another slave, and they killed him; and they treated many others in the same way, beating some and killing others. The only one left to send was the man's own dear son. Last of all, then, he sent his son to the tenants. "I am sure they will respect my son," he said. But those tenants said to one another, "This is the owner s son. Come on, let's kill him, and the property will be ours!" So they grabbed the son and killed him and threw his body out of the vineyard. What, then, will the owner of the vineyard do?" asked Jesus. "He will come and kill those men and turn the vineyard over to other tenants."
Mark 12: 1 - 9.

The vineyard is our land and forests.

Pastor Simon explained this story to the people. He said that Jesus told this parable to describe how the Jewish leaders were treating him and what they would do to him. "But," he said, "the parable also has another meaning for us today.

"The man who planted the vineyard is God. The vineyard is the land that provides food for us and many other things we need. The vineyard includes our towns and villages, the forest, the rivers and the springs of water. We are like the people in the story who rented the vineyard. The vineyard is God's, and that means that the land we live on and cultivate is God's. He has put us in charge of it to take good care of it.

"In Jesus' parable, the man himself put a wall around the vineyard, dug a hole for the wine press, and built the watchtower. These things were what the people needed to care for the vineyard, to protect it, and to get a good harvest from it.

"God provides what we need to take care of the land he has given us. This includes good minds to think, to study, and to find the knowledge we need to care for the land. We have strong bodies to work hard. He has given us families, friends and communities. We must work hard to care for what God has made and given to us."

He paused in his message and looked out the window of the church building. "Years ago we had big forests on the hills and even near the town. But we have cut most of them down. The spring of water at the base of that hill over there where the forest stood has dried up. The soil in our fields has become poor and produces little food. We have not taken good care of the 'vineyard' God has given us. I am going to invite you to come to my house this afternoon so we can talk about this together."

That Sunday afternoon a group of people from the church came to Pastor Simon's house to talk. The ideas in his message were new to them. Some of the people did not like what he had said. They thought he was saying they had done wrong. They all had questions they wanted to ask to learn more about this.

Discussing the problem of poor land.

David, the farmer, spoke up. "My fields belong to me. The head of the village gave them to my grandfather and we have farmed them ever since. Why are you saying the land belongs to God?"

Everything belongs to God

Pastor Simon relied, "The land, the fields, the forest, and the springs of water belong to God because he made them. They do not belong to us. God said to his people, 'The whole earth is mine.' (Exodus 19:5) We should not say, 'This is our forest; these are our fields.' Rather we should say, 'This is the forest and these are the fields God has given us to use.' So we must take good care of the earth and everything on it that God has made. One day when we come into God's presence, he will ask us how we have cared for the land he has given us.

"We must not take what God has given us and try to make it our own. We must never say, 'This is mine and I will do what I want with it. No one can stop me.' That is not true. If we use in a bad way something God has given us, he will take it from us. He will give it to someone else as in the story Jesus told."

Anna, David's wife, said, "You say that God took land away from people because they had not used it well. Where do you see that in the Bible?"

"Anna," Pastor Simon said. "God gave the land of Palestine to the people of Israel. They did not use it as God wanted them to. So he drove them out of the land and gave it to others. You see, God gives us the wonderful privilege of using what he has made. But he expects us to use it wisely and not destroy it. We have destroyed the forest he gave us. We also seem to be destroying the soil that grows our food."

We can use what God has given us

Roger asked, "If the land is God's, how dare we use it to grow food for ourselves?"

"Roger, God is the first farmer. He planted a wonderful garden, with many different kinds of trees, plants, and flowers. He created Adam and Eve, put them in the garden, and told them to cultivate it and grow their food there. He also told them to take good care of it. He wants us to do the same today. God wants us to use what he has given us to make our lives good. We can eat the food the land produces. We can drink water from the springs or wells. We can use the trees of the forest for the things we need.

"However, we must not destroy what God has made. We must use everything wisely and care for it well. God sees what we do. If we work hard and take good care of what he gives us, he will bless our work, and we will live well. If we do not take care of what he gives us but destroy it, we will lose it. He will take it away from us and give it to someone else."

The forest is gone

Pastor Simon stood up and pointed to the hill to the south. He said, "Many years ago, on that hill, our people had a big forest. They had many trees for building houses. Animals lived in the forest. They got much food from the forest–fruits of many kinds, mushrooms, and the meat of animals.

"Now the forest is gone. We have cut it all down. We wanted to have big fields to grow more food so we could sell it and make money. We cut down trees to sell to other people for making houses and furniture. We did not let the trees regrow or plant new ones to replace them. The forest is now gone."

Arthur added, "Pastor you are right. We now have no trees for building houses. We have much trouble finding firewood and must pay money for it. It is true that the spring of water at the bottom of the hill has dried up. Is that because the forest is gone?"

Rebecca spoke up. "Some people say God is angry with us and took the spring away from us. Is that so?"

"No, I don't think so," the Pastor replied. "In Genesis chapter two God told Adam that what we do has consequences. The forest lets rain water go down into the ground and into the spring. Our ancestors knew that and they protected the forest. When we cut down the forest, the rain water runs off and is gone. It can no longer go down into the ground and into the spring. The dried-up spring is our fault, not God's fault. Furthermore, as the rain water runs down the hillside, it has made those big gullies that have taken away more good land from us."

Robert, the school director, said to Pastor Simon, "It was our ancestors who gave us our forest. Why did you say it was God who gave it to us?"

Before the pastor could reply, Yvonne, the nurse, spoke up. "Robert, who gave the forest to our ancestors? It was God, wasn't it? Our ancestors knew it came from God. They knew they must take care of it. They passed the forest on to us to take care of it. But we have cut down all the trees for wood and to make gardens. So the forest God gave our ancestors is gone. I am sure God is displeased with us, and certainly the ancestors would also be displeased."

Pastor Simon said, "What Yvonne says is right. So now we must talk about what we should do."

Should we plant trees?

Robert asked, "Are you saying we must go out and plant trees?

Pastor Simon asked, "What do the rest of you think?"

Roger, a carpenter, said, "Planting trees is not our work. God is the one who plants trees, not man. If God gave us the forest, we must ask him to give it back to us. That is God's work, not ours."

Pastor Simon spoke again. "Roger, it is true that the Lord God planted a garden called Eden. He made all kinds of trees grow out of the ground. The trees were beautiful and some gave food to eat. God put man in the garden and told him to take care of it."

Roger said immediately, "That is exactly what I just said. God is the one who plants trees, not us."

Pastor Simon replied, "Roger, remember that God said, 'Let us make man in our image.' So God made man in his image, man and woman. We have God's image in us, the image of a God who plants trees. Therefore, God made us to be tree-planters just like him.

"Listen to what God then said to Adam. 'I am giving you all these good things so that you can use them. You are to work hard to care for them and to protect them' Genesis 2: 15.

"You see, God took much time and effort to create all these different kinds of trees. He did it to help us, so that we could have much food and wood and have the beauty of the forest for our pleasure. He then made us to work with him in caring for the land and the trees. He gave us the responsibility to take care of all he had made, not to destroy them. If we have cut down all the trees God planted, how can we ask him to replant them for us? We must do that ourselves."

Roger was a bit upset. "You mean we can never cut down trees? That's ridiculous."

The Pastor smiled. "We can cut down trees and use them to build houses and other things. That's why God gave them to us, to use to meet our needs. But when we do cut them down, we must plant new trees to take their place. That is how we should take care of the forest. In this way, it will not disappear."

The forest and the birds

An old man who had many children and grandchildren spoke up. "Pastor Simon, I have observed that we do not have many birds now. Many years ago, when I was young, we had lots of birds around our homes, in the fields, and in the forest. I loved hearing them sing. But now I seldom hear birds sing. Have the birds disappeared because the forest is gone?"

"Exactly," Pastor Simon replied. "The birds make their nests in the trees. When the trees are gone, the birds no longer have a place to build their nests. So they go away as well."

Arthur added, "I don't see the big birds of prey, like hawks, flying around any more. That also may be because the forest is gone."

"Yes," Pastor Simon said. "Those birds feed on the many small animals that live in the forest and under the trees. Now that the forest is gone, those small animals are gone. So the birds of prey have gone somewhere else. They no longer eat the field rats, mice, and insects that eat our grain and damage our crops in the field."

Protecting trees and gardens

David spoke up again. "Pastor, two years ago I planted some mango trees and two guava trees. They started to grow well. But then my neighbor's goats came and ate them up. What can I do?"

"That is why the man in Jesus' story who planted the vineyard put a wall around it," the pastor replied. "The wall was to protect it from wandering animals. Here is a serious problem we must talk about. We let animals go where they wish. They get into our fields and gardens and destroy the food we need for ourselves. We allow animals to rule the land. Yet, God told us we are to rule over both the land and the animals. It is urgent that we find ways to do that and not let animals rule us or the land."

Arthur spoke up again. "Pastor, last year two-legged animals got into my garden and took half my groundnut crop."

Yvonne said to him, "You mean thieves stole part of your crop?"

Arthur replied, "Of course, that's what I said. What can I do?"

Pastor Simon said, "That is why the man built a tower in the vineyard. A watchman stayed on that tower when the grapes were ripe to protect the vineyard against thieves. We must talk about ways to protect our fields from thieves."

"Better yet,"Yvonne said. "If we live as God wants us to live, there will be fewer thieves."

Stop destroying trees and start restoring the land

After much discussion the people decided they must do something about the land and the forest. Pastor Simon spoke again, "We have something very important to do first." He opened his Bible to the Second Book of Chronicles and read chapter 7, verse 14:

> *If my people pray to me and repent and turn away from the evil they have been doing, then will I hear them in heaven, forgive their sins, and make their land prosperous again.*

"If we want God to help us, we must first repent of the bad things we have done to the land on which we live. Repent means two things. First of all, it means to turn around. It then means to go in the opposite direction. We must first turn away from destroying our forest and trees, and from misusing the land. We must then begin to restore the trees and the soil, and improve the ways we grow our food."

Anna spoke up again. "Pastor, let's make a list of the things we need to repent of and what we need to do about our restoring our land."

Pastor Simon smiled. "Good for you, Anna. Let's all suggest some of the things we must do. Rebecca, please write down these suggestions on some paper."

The group suggested many things, and this is what Rebecca wrote down.

1. We have cut down much of our forest. We must now let the forest come back and grow well. We may need to plant forest trees ourselves to help the forest return.
2. We have cut down many fruit trees. We need to replant them and even new kinds that can give us food.
3. We have planted too many fields and gardens and not given the soil enough rest. So it has become infertile. We must find ways to make the soil better so that it can produce more food.
4. We have to find better ways to raise animals and keep them from destroying gardens and fields.
5. Every year we burn large areas of grassland and destroy the plants that cover and protect the land. We must discuss when to burn the fields and why not to burn the grasslands.
6. We have allowed the rains to wash away much good soil from the land. Now we must see how to keep erosion from destroying the fields.

Pastor Simon said this was all good. These were things they needed to repent of, stop doing, and then start working together to do the right things. He said, "God has showed us some good things today and we must follow this.

"We can also learn something from the example of our ancestors. Before they cleared a field or planted it, they prayed, because they knew they needed God's wisdom. We must do the same. When they harvested their food crops, they thanked God for them. We must remember to do that as well because all good things come from God."

They all agreed and began to pray. They confessed to God how they were misusing the land, the forest, and the soil. They asked for God's forgiveness. They then asked God to give them wisdom how to restore the land and the forest and make them prosperous again. They prayed and said, "Dear Father in heaven, please show us all that we must do."

God did give them wisdom. During the next few years they followed what they learned from God and from people who had much knowledge about the land and forests. They now have a fine forest, with trees they can use for building, wood for fires, and fruits to eat. Many animals and birds now live in their forest. God gave them wisdom about how to help the land become stronger and grow more food. Pastor Simon and the people of his church and community then began to help people of other communities find ways to care for the land, forest, and springs of water God has given them and how to grow more food.

<u>Questions for study</u>

1. *Read carefully Genesis chapter 3.*

- *Why did the woman and the man obey Satan and not God?*
- *What happened to their relationships with God, with each other, and with the land?*
- *How does disobedience to God make the soil infertile?*

2. *Remember this: the curse of God and the curse of man are different.*
- *Man gives a curse to harm or destroy something or someone.*
- *God gives a curse as a warning that what we are doing is wrong and will have evil results.*

3. *Why is God concerned about how we use the land and forest that we have?*

4. *What happens when we do not take care of these wonderful resources?*

5. *When the soil becomes tired or the spring of water dries up, is that God's fault, or ours? How do we know that?*

6. *Why do the birds and animals go away when the forest disappears?*

7. *Make a list of the problems of our own land and forest.*

8. *Make a second list of what we must do to restore them.*

9. *Is farming a respectable way of life?*
- *Why do you think it is?*

10. *The Bible says we are to glorify God in all that we do – I Corinthians 10:31.*
- *How can we glorify God through farming?*

Chapter 2

Caring for the Fields

Pastor Simon was happy. His nephew John was visiting him for several weeks. When John was six years old, his father had died. Although Pastor Simon had little money and three children of his own, he had paid the school fees for his nephew. John was now studying in an agricultural school in the city, and he was on vacation.

One evening, John explained to Pastor Simon what he was learning at school. He told about the wonderful developments in agriculture in Europe and North America. The farmers there grow much food, and no one goes hungry. John's teachers had studied agriculture in Europe and were teaching the European methods of agriculture in John's school.

"Many years ago," John said, "our forefathers knew how to grow enough food for our families. The land was quite fertile and gave good crops. They had plenty of land because the population was not big. Each year they would clear the land they needed and plant their crops. They could then let that land rest for five or six years so it could become fertile again."

Cutting down all trees before planting

"In more recent years, however, the population has grown, and it was no longer possible to let fields rest for five years. The land became poor and could no longer produce good crops. We then learned how people grow food on farms in Europe. The farmers cut down all the trees so they can use much land. It is very important for each field to be free of all trees and bushes. In that way the crops grow better and produce much food. So we are teaching our people how to cut down all the trees before planting."

Pastor Simon was silent. It was true the fields produced little food in recent years. The people in his town of Katindi had learned from the Europeans how to clear the fields completely before planting. But after a few years they discovered that this does not help the fields grow more food. They saw that the hot sun burns the crops and dries out the soil. So they began looking for better ways to grow food.

After a few minutes Pastor Simon said, "Yes, John, I understand what you are saying. Tomorrow let's walk to the village of Malenga. You will see how the people there have cut down all the trees to make fields for growing food."

"Do they have much food now?" John asked.

"You will see tomorrow," Pastor Simon replied.

Cutting down all trees causes problems

The next day they walked to Malenga. They went to the house of the head man. "Moses," Pastor Simon said. "This is my nephew John. He is studying agriculture in the city. We came to see how your fields are doing. Did you have a good harvest this year?"

The head man shook his head. "No, Pastor. Our harvest was poor and we soon will be hungry. We have an agricultural agent here, and we have been doing what he taught us. Each year we clear the land carefully and burn all the trees and bushes. We make the soil smooth before we plant the seeds. But the land seems to be tired out. The soil is now mostly sand. It is very dusty. When the wind blows, it blows away a lot of soil. Our millet and maize produce little food. Even the cassava and groundnuts grow poorly. We don't know why."

Together they walked through the fields of the people of the village. They saw that almost all the trees were gone. The soil did indeed look poor. In many places it was very sandy. The men could see that crops could not grow well there. They saw some places where the land had become hard like cement. In those places it was impossible to cultivate. They did see one kind of weed growing in many places, but not even the animals would eat this weed.

Poor fields yield poor crops.

"Pastor Simon, what are we doing wrong?" the head man asked.

"Moses, tomorrow come to our town. You will see how we are doing things differently in our fields. We can talk about this problem there."

When Moses, the head man of Malenga, arrived at Pastor Simon's house, they sat under a mango tree and drank tea. Pastor Simon explained that years ago agricultural teachers had come to their town and taught the people how to clear the land before planting their crops. The people had objected because that was not how the ancestors had cultivated the fields. The teachers explained that these new ideas came from Europe where farmers grow much food. So the people of Katindi followed the instructions of the agricultural agents. But soon they saw the soil was becoming weak and little food would grow. The people themselves met together to discuss the problem.

The people knew that the traditional methods had produced food but now, with more people to feed, it was not enough. The European method of clearing the land was producing even less food. They did know they must cut down the trees and bushes from some of the land in order to plant their crops. But should they cut down everything?

Pastor Simon said, "I suggested we see if the Bible could help us. We read the story of creation in Genesis chapters one and two. God made the land and the soil, and everything that grows from the soil. It was all good. Then we read this:

> *When the Lord God made the universe, there were no plants on the earth and no seeds had sprouted, because the Lord God had not sent any rain, and there was no one to cultivate the land. (Genesis 2 : 4 and 5)*

"We asked what this meant. What was God trying to teach us? Is there a relation between the lack of trees and plants and the absence of rain? We saw that trees and plants cannot exist without rain, and that rain does not exist when there are no trees or plants. When that condition exists, man cannot till the ground. We concluded that cutting down all the trees is not good. We should leave some of the trees to protect the land. We must also keep some of the forest so rain can continue to fall. We saw how God brought together trees, plants, rain, and people in the Garden of Eden.

Farming God's Way

"God told Adam and Eve to use the land, cultivate it, and take care of it. God said to them, 'Take charge of the animals, the plants, the land and rule over them.' Let me ask you a question, Moses. If you want to use something wisely and rule over it, what must you do first?"

Moses thought for a short while. He was the head man of his community, the ruler of the village. He could rule well because he knew his people, how they lived, their land, and where they get water. So he replied, "To rule my village, I have to know my people well and care for their needs."

"Exactly," said Pastor Simon. "How did you come to know your people well?"

"I grew up among them," said Moses. "I got to know them all and I learned about each one."

Pastor Simon said, "That is what we must do about the soil that grows our food. We must study it to find out what makes it produce good food and how we must care for the soil."

"How did you do that?" Moses asked.

"In Europe the people studied their land. They used science to find out how to help the land there produce much food. One method that helped the land produce food in Europe was to clear all the trees from the fields before planting. They could then plow the soil deeply, plant it, and have crops grow well."

Moses spoke up. "But Pastor, this is Africa. What works in Europe is fine, but would that work here in Africa? Isn't Africa different from Europe?"

Pastor Simon smiled, and John was listening closely. "That is exactly what we talked about. In Europe the sun is not very hot when planting time comes. It does not burn the soil there. So they cut down all the trees and plow the soil deeply.

"Here in Africa the sun is very hot. It dries the soil quickly if the soil is not protected by trees, bushes, or plants. During the rainy season, we have much rain. It falls on the ground very hard. When there are no plants, trees, or bushes to protect the soil, the heavy rain can take away much of the soil. And when there are no trees, the wind blows away a lot of soil.

"We studied the book of Genesis. We saw that God had planted trees everywhere, and they covered the soil. We discovered that soil in the forest is rich. The trees cover the soil and protect it. When you put your hand down into forest soil, it is soft and cool. That is because the temperature in the forest is cool. Leaves fall from the trees and lie on the soil and they rot slowly. They go down into the soil to make it rich.

"When the soil has no cover, the hot sun quickly destroys the fallen leaves. They cannot add rich food to the soil. The soil becomes hard and hot. Crops do not grow well."

"So what did you conclude?" Moses asked.

No deep plowing of the soil

"We concluded that it is good to keep the soil covered at all times. We also decided not to plow the soil deeply and expose it to the sun."

Moses asked, "You mean turning the soil over by plowing it is not good?"

"Exactly," Simon replied. "That exposes the good things in the soil to the hot sun. The sun burns them up quickly and they are no longer there to help the new crops grow."

"How can we plant our crops if we do not plow the soil?" Moses asked.

Pastor Simon explained, "In our fields, we cut down the weeds, grass, and stalks of the old crop. We do not burn them but let them dry on the ground. We spread them out over the whole field. These dried plants decompose slowly and protect the soil from the hot sun. They keep the soil cooler so it does not dry out. They also protect the soil from erosion by heavy rain or wind."

"How do you plant your seeds when the ground is covered with that grass and weeds?" Moses asked.

"We use a hoe to loosen the soil where we want to put in the seeds. We then make a little hole and plant the maize, millet, or groundnuts in the hole. We leave the mulch between where we plant the seeds. We try to cover the whole field with mulch like a blanket covers a bed. This helps stop weeds growing in the field."

"Do the little plants grow well?" Moses asked.

"Oh yes," Simon replied. "The soil is loose but covered and protected from the sun and heavy rain. The blanket of mulch slowly decomposes and provides food for the young plants. So they grow very well."

The pastor continued. "So we decided it is not good to cut down all the trees. We now have trees growing around our fields and a few in the fields. They protect the soil and the crops from the hot sun and the heavy rain."

"I want to see what you do here." Moses said. "I want to know, because my people cannot grow enough food. We may have to leave our village and our burial grounds and go somewhere else. Did you plant new trees?"

Planting new trees to protect fields

Pastor Simon replied, "In many fields we planted trees like acacia trees that give some shade but not too much. We also planted new trees around our fields. We will go see them shortly. In a few of the fields we have cultivated only recently, it was not necessary to plant new trees. We discovered that the roots of the trees we had cut down were still alive in the ground. Each year they would try to grow again, but we kept cutting off the young shoots. So we decided to let some of the old trees grow again.

"When we began to prepare the fields, we found many shoots growing from these old roots. We cut off some of the shoots and let others grow. That way the land would have some protection from the sun and from the heavy rain and the wind. We let different kinds of trees grow again. We knew God had made many different kinds of trees. He must have had a good reason to do that. In the forest we see many different types of trees all growing together. So we decided to let trees of different kinds grow back in our fields, trees that would make the soil stronger and that would give some shade, but not too much."

"Can we go see what happened?" John asked.

The three men walked out into the fields west of Katindi. They could see fields of millet and maize, and others of ground nuts and cassava. In some places vegetables were growing. A few big trees were growing in the fields, but not close together. Rows of trees grew along the borders of the fields.

Field with trees and good crops.

"Since these trees have grown to this size, we do not have bad dust storms," Pastor Simon said. "In the past, when we cleared the fields, the wind would blow much of the soil away. Now that does not happen any more. The heavy rains do not wash away the soil. The trees protect the soil from the strong wind and the heavy rainfall. So the land is better."

John asked, "Does the shade from the trees keep the crops from growing under the trees?"

"Not really,"Pastor Simon replied. "Look at these big trees in the fields. Many of them are acacia trees. Their leaves are small and high up. Enough sunlight gets through to the ground so the crops can grow. These acacia trees also add minerals to the soil and make it stronger. The temperature in the shade of those trees is cooler and that helps the crops grow better. Another advantage is the leaves that the trees drop on the soil. They soon decompose and become food the new crops can use to grow better. Other kinds of trees also let a lot of sunshine through and protect the soil. Those are all good trees to leave in the fields."

"But we must get all the crops we possibly can," said Moses. "That is why we cut down all the trees so there will be no shade at all."

Pastor Simon laughed. "Moses, that is why the hot sun burns your soil and destroys the things in the soil your crops need to grow well. Here, these trees protect our soil from the hot sun and the hard rain. The rain must go through the branches of those trees and it then comes down on the soil gently. We now get much food from these fields. In the past we only got a little food."

"How can this be?" asked John. "Why don't my teachers know this?"

Pastor Simon shook his head and looked a bit sad. "John, your teachers studied agriculture in Europe. Have any of them gone out to the villages to see the problems?"

"Yes, and we go with them. They go to teach the people and we help them. They work with the people as they are putting these new methods into practice. They know what is going on."

Pastor Simon asked, "Do the teachers ask the people questions? Do they try to find out the traditional methods of agriculture? Do they listen to the people? Do they try to discover the wisdom of the people?"

"Oh no," John said. "They think they know the right methods because the methods come from Europe. So they go out to teach them to the people because the people do not know those new methods."

Moses spat on the ground. "Sure. We did what they taught us, and now we are hungry. European methods don't grow much food here in Africa."

John spoke up. "I think I am beginning to understand. Our ancestors did have a hard time growing food. They often were hungry, because they did not know how to grow food well. The Europeans have found many good methods for Europe. But their methods do not all work well here in Africa. Uncle Simon, you and your people went to God. You studied the Bible. You took ideas from the Bible and some from the ancestors. You learned some good things from the Europeans. Most important, you studied the situation and used the wisdom God gave you to see what works. Now you have enough food."

Pastor Simon was pleased. "You have spoken well, John."

John said, "Thanks, Uncle Simon. But I have another question. You talk about the Bible like it was a textbook of agriculture. I don't find any real agriculture in the Bible. So what are you saying?"

"John, that question calls for another cup of tea. Let's go back to the house and sit in the shade. Moses, would you like to join us?"

"I would indeed," Moses replied. So the three men walked back to Pastor Simon's house.

Questions for study

1. *What does God say in the Bible about the relationship between plants, trees, rain, and man? See Genesis 2 : 5, 6*

2. *What happens when we cut down all the trees from our fields?*

3. *How does deep plowing hurt the soil?*

4. *Why does having some trees in the fields help the crops grow better?*

5. *What are good trees to let grow in the fields?*

6. *What was the problem with John's agricultural teachers?*

Chapter 3

Finding God's Laws Through Science

The late afternoon sun was shining through the leucaena trees as the men sat drinking tea in back of Pastor Simon's house. Moses commented about how pleasant his back yard was -shady, cool, and with many flowers planted around the yard. The pastor explained that his wife Virginia loved flowers. He said, "I believe God took extra care in designing the many varieties of flowers so that we would appreciate the different colors and give thanks to God for them."

Simon, Moses and John having a serious discussion.

"Uncle Simon," John said. "You say you have gotten wisdom from God, the Bible, and science. Can you explain this to us?"

The Bible and science

The pastor smiled. He said that many people today believe the Bible and science do not agree. That is because they do not fully understand either the Bible or science itself. In reality, the foundation of science is in the Bible because science studies what God has made. The laws science discovers are laws God has put into all that he has created.

"For example," he said. "The story of creation in Genesis chapter one shows the marvelous order God created in nature. He made the sun, moon, and stars and put each one in its precise location. Tell me, John, where the sun comes up each morning."

John pointed to the east. "Over there," he said. "But sometimes it is more to the south, and other times to the north. How can you say that is precise?"

"The sun does rise in different places at different times of the year," the pastor replied. "That is because the earth is tilted on its axis. We study carefully the earth and how it goes around the sun once a year. We know the angle at which the axis of the earth tilts, and how it spins on its axis precisely every 24 hours. In this way we can then know exactly at what time and at what place the sun will rise every day of the year. It will never change, and has never changed for thousands of years. God is a God of order and of precision."

God put laws into all of nature

"What does this have to do with growing food?" Moses asked.

"Moses," the pastor answered. "If nature did not have a precise order scientific study would be useless. There would be no laws by which all things operate. We would have no way of knowing how to grow lots of food. However, the world God has made is orderly. All the heavenly bodies move in precise ways. So we can study them. God made the human body in a precise way. Doctors study the human body. So they know how to help us maintain our health, what foods we should eat, and how to heal us when we get sick. When we discover the laws God has put into everything – the soil, the crops, the climate – we can follow them. By following these laws of nature, which are God's laws, we can benefit from them."

The pastor added, "We also need to understand that God is always on time. The days, the seasons, the years all come and go precisely on time. That is because God has made time, and we must respect that. We must know the best time to plant, to harvest, and to make the best use of the time God gives us."

"Are the same things true for agriculture?" John asked.

The laws of agriculture come from God

"Of course," the pastor replied. "You are studying the different kinds of soil God has made. Some soils are sandy. Others have much clay. You also study different plants and crops. You know which ones will grow well in sandy soil, and which will grow well in soil with clay."

"What does that have to do with God?" John asked.

"John, stop and think. God made each crop to grow in a particular soil. By studying each crop you discover in which kind of soil God has prepared it to grow. You plant it there, it grows well, and you get lots of food. But if you don't know those laws God has made, how are you going to grow lots of food?"

"Uncle Simon, are you saying that the laws of agriculture we discover have come from God?"

"Exactly. They are the laws God has put into the different soils, into the trees, into maize, millet, and all the foods we grow. If we know these laws that God made and respect them, we can grow the food we need."

"That's amazing," John said. "I never thought that agricultural science came from God."

Discovering God's laws

"You see, John, we can discover God's laws in two ways. One is by studying the Bible. It shows us many important laws about how we are to live. However, the Bible says very little about agriculture, nutrition, hygiene, and other practical matters. To learn those laws, we use science. By scientific study we see how God has made all things. The laws we discover by the scientific method have actually come from God."

"Do the Europeans know that?" John asked.

"Some do. Some don't. Even if they do not know the origin of the laws, the laws remain the same. However, the problem here in Africa is this. The Europeans, by their scientific studies, discovered how to grow crops well in their conditions. The mistake many of them made is to believe that those crops would grow well in Africa following the laws they discovered in Europe. However, as we have seen, what works in Europe does not always work in Africa because conditions here are different. We must study our own conditions here – the soil, the climate, temperature, rainfall, and so on."

"Uncle Simon, I am beginning to understand what you are saying. Even though you have not studied agriculture, you have observed conditions here. You have seen what happens to the soil and the land when certain things are done. You have seen that some things help the soil; other things hurt the soil. And you concluded that this is because of the way God has made things."

"You have spoken well, John. Now let's look at an important chapter in the Bible. It is chapter 28 of Deuteronomy. Here God is telling his people about the importance of all of the laws he has given them. If people obey his laws, they live well. If they do not obey his laws, they will have many difficulties, even many illnesses. God does not send these difficulties or illnesses. They come because of our disobedience."

"Uncle Simon, it says here that if the people disobey God's laws, he will curse them. Why does God curse people?"

God's "curse" is his warning to us

Again Pastor Simon smiled. "John, we must understand the great difference between the curse of man and the "curse" of God. The curse of man is to harm or destroy. The curse of God is to warn us that we are doing wrong. So when we see the soil is no longer fertile, or the spring of water has dried up, we should not blame God. We should recognize, however, that these things are warning us that we have not obeyed God's laws. Therefore we need to discover the laws about the soil and springs of water and follow them. So in reality, what the Old Testament calls the "curse of God" is simply the bad results of our failure to follow the laws God has put into everything he has made."

Moses shook his head. "Pastor, all this new stuff is too much for me. I just do what my ancestors always did."

The pastor laughed and said, "Moses, actually our ancestors knew a great deal about nature. They knew it came from God. They knew exactly where to plant maize, and cassava, and groundnuts. They planted them where they knew God had prepared those crops to grow well. They also knew God has given us the forest to benefit us. They protected the forest and the soil because they knew God wanted us to do that."

"So why do we have all these problems now?" Moses asked.

"The answer to that is simple," the pastor replied. "Many things have changed, but we have been slow in changing our ways. Change is difficult, yet necessary. Before we make big changes, we must think carefully, study the situation scientifically, read God's Word, and ask God for his wisdom. Moses, the hour is late and the sun is about to set. Spend the night with us here. We will have dinner, and then continue our discussion."

"Thank you, Pastor," Moses replied. "I will do that because I have many more questions to ask."

Questions for study

Read carefully the whole chapter 28 of Deuteronomy.

1. *What benefits come if we learn and obey God's laws?*

2. *What are the bad things that can happen if we do not know or obey God s laws?*

Chapter 4

Dominion that Restores and Improves the Land

That evening Pastor Simon had a fire in front of his house. Men and women from the town came to talk. Moses was there because he had decided to spend the night with Pastor Simon. He still had many questions to ask. Pastor Simon read from the Bible.

> *Those who wept as they went out carrying the seed will come back singing with joy, as they bring in the harvest. (Psalm 126: 6)*

Weeping when planting seeds

He said, "The man who wrote that Psalm knew our situation very well. When the harvest came we would rejoice, and our children would dance and sing. We knew we had food to eat. We could eat two meals a day and our hunger would go away."

Arthur spoke up quickly, "Yes, Pastor, but after three months the ground nuts were gone. The supply of millet and maize would diminish, and we could only eat one meal a day. Our hunger would come back. Often we would not even have enough cassava to eat."

Rose said, "Pastor, when I was a little girl, at planting time we could eat only one small meal a day because the food was almost gone. I was always hungry. One day, however, I went into the little storehouse in back of our house and my eyes grew very big. There was food there, a lot of food. There were many stalks of millet and a big bag of maize. I ran to my father and shouted, 'Food, Papa, food. We have lots of food. Let's eat.'

"My father took me in his arms and said, 'Rose, what you saw in the storehouse is not food for us now. It is food for us next year. Those are the seeds we must soon plant in the ground so we can have something to eat next year. If we eat it now, next year we will have no food. We will die of hunger.'

"A few days later, my father and mother took the millet and maize, and we all went to the fields to plant. I cried and cried to see that food go into the soil because I was so hungry. Like the Psalm said, 'We went out weeping to plant seeds.'"

Rose weeping as her parents plant seeds.

Pastor Simon said, "Yes, Rose, that is what the Psalm means. We must control our hunger now so that we can eat later. We must think about what we will need tomorrow, and not just about what we need today."

Moses scratched his head and then said, "I understand that Psalm very well. Part of the year we have plenty of food. Part of the year we are hungry. But Pastor, how is it that in your village you have food that lasts all year long?"

Pastor Simon replied, "Moses, this morning I showed you some of the things we have done. We let some trees grow to protect the soil. This helps the crops grow. Tomorrow I will show you more things we have found by studying the land, the trees, and the different foods we eat. Some of these things we have learned from the Europeans. Others we have learned from our own experience."

Roger then said, "Moses, the pastor helped us see how we could put together the wisdom of the Bible, the wisdom of the ancestors, and the wisdom from Europe. Now we have plenty of food. Isn't that wonderful?"

They all agreed, and Moses himself could see that the people of Katindi were eating well. John could see this also.

Telling others about God's Word

Pastor Simon went on. "God is saying something else to us in the Psalm. The Bible often compares seed to God's Word. We are the ones who should plant this seed. We plant this seed when we tell others about Christ and what God wants us to do.

"It is not always easy to do that. Many people do not want to follow God's Word. They do not want to follow God or obey his commandments. Often we weep when people refuse to accept God's Word. But we must keep on planting it. We must keep on telling other people about Christ and how he can give us new life.

"This Psalm is saying that if we do this faithfully, one day we will rejoice. Some people will accept Christ and receive God's Word. They will find real life and a better life now. That is when we will rejoice with them."

Moses said, "Pastor Simon, you say that God shows us how to take care of the land and how to plant our seed to have good food. Does he also want us to tell others what he says?"

"Yes, Moses, that is the work God gives us to do. He wants us to work hard for ourselves. He also wants us to help other people. That is why we came to your village and then invited you here to see what we are doing."

Having dominion over the land

John said, "Uncle Simon, you said God told Adam and Eve to take charge of the land and have dominion over it. But they sinned and disobeyed God. Did they lose dominion over the land, the animals, and the soil?"

"That is a very important question, John. Would anyone here like to try to answer John's question?"

Virginia, Pastor Simon's wife, said, "Pastor, let me try."

"Go ahead, Virginia. We are listening."

"I think Adam and Eve did lose their dominion over the land and the animals. God told them to rule over nature with him. As long as they lived with God and got their wisdom from him, they had the right kind of dominion over nature that pleased God. They had plenty to eat and the animals did not harm them.

"You remember, however, that God told them not to eat the fruit of the tree of the knowledge of good and evil. If they ate that fruit, they would die. Then Satan came to them in the form of a reptile and tempted them to eat that fruit. Satan wanted them to decide to get wisdom from him rather than from God.

"When they accepted Satan's wisdom, they sinned against God. They refused to follow God's wisdom and get help only from him. They lost much of the dominion over nature God had given them. They also misunderstood the dominion left to them. They thought it meant using nature as they see fit, selfishly for their own benefit."

Adam and Eve eating the forbidden fruit.

Taking responsibility for the land

Robert joined in. "That is a good thought, Virginia. However, if we lost most of our **dominion** because we sinned, we did not lose our **responsibility** for nature, for the animals, and for the land."

Pastor Simon said, "You have spoken words of wisdom, Robert. God has given us the land and we are responsible for it in spite of our sin. We are responsible to God for how we use what he has given us. We are to use it for the good of others and not just ourselves. We are also responsible to ourselves and to our children. If we destroy our land, the forest, and the animals, what will our children eat? How will they live?"

Moses said, "I am still having trouble understanding this. Please explain it to me better so I can understand."

Pastor Simon began, "God created all of nature – the land and water, plants, animals, mountains, rivers, trees and grasslands. Everything God created is good. He put everything in its proper place – the sun, moon, and stars; the mountains and valleys, the rivers and streams, the trees and the food we are to eat. He created many different kinds of animals. Each kind, or species, is different from every other kind. Everything follows the order God has established.

"God created Adam and Eve and put them in the midst of nature. He breathed his spirit into them and put his image in them. He told them to rule over the nature he had created. He wanted them to rule over it with him in the right way, like a son who learns from his father. The wind, the rains, the storms were all good and man was not afraid of them.

Moses said, "God also created the animals. Did he put his spirit in them?"

Pastor Simon replied, "Let's read how God created man and the animals. In Genesis 2: 7 we read that God formed man out of the dust of the ground and breathed his spirit into him. Man then became a living person. We are spiritual beings, and in this way we are like God.

"Genesis 2: 19 says that God formed all of the animals out of the dust of the ground. He brought them all to Adam so he could rule over them. It does not say that God breathed his spirit into them. So they are not spiritual beings like we are."

Moses said,"Many people believe that animals, like lions, elephants, snakes, and many birds have powerful spirits that can harm us. Is that so?"

"No, it is not so," Pastor Simon answered. "Animals have physical bodies like we do. With their small brains, they can feel certain things, but they don't have a spirit. When God made them, he told Adam to name them. When you name something, you have a certain power over it. God wanted man to have power over the animals, and not for the animals to have power over man."

Moses asked, "But how can we have power over lions, snakes, and other animals and birds?"

Pastor Simon explained this to him. "When man sinned by refusing God's will, he rejected the order God had established. So disorder came into the world. It came even into nature and nature became disorderly. Man became afraid of nature – animals and birds, the wind, rain, storms, lightning, and fire. He worships things in nature instead of worshiping God. Instead of trusting God to protect him, man offers sacrifices to things in nature, praying to them to ask them not to harm him. The Apostle Paul said in Romans 1: 25: 'They exchanged the truth of God for a lie, and worshiped and served created things rather than the Creator.'"

Moses was amazed, and said, "Pastor, we always pray to the spirits of our ancestors before we plant our fields or harvest our crops. We pray to the spirit of the rain to send rain at the right time. We pray to the spirit of the soil to give us much food. Are you saying that this is not good?"

Trusting in God alone

Pastor Simon answered, "Moses, God wants us to trust in him alone. We are not to pray to any other spirits or worship any other spirits. We are to pray to God, however. Just like you have always done, we should pray before we plant and before we harvest, but praying to God and not to the ancestors or other spirits. If we trust in God, understand what he has said in his Word, learn from what he has created, and apply this, our lives will be much better.

"You see, God wants us to rule over nature and not to be afraid of it or worship it. He wants us to rule over it in away that restores and improves the land. We can rule over nature only with God's wisdom and help. In this way we can restore order and have control over some of nature. We can never control all of nature because sin brought much disorder even into it. But we can rule over the trees and the forest, the land, the soil and many of the animals and make them better. We are even to rule over our own animals."

Ruling over the animals

"I sure don't understand that," Moses said. "Our cattle, goats, and sheep go where they want to go. How are we supposed to rule over them?"

"By keeping them in their proper place," the pastor replied. "Here in Katindi we have had to work hard to accomplish that and we have succeeded fairly well. After you have seen what we have done in our fields and forest, we will talk about ruling over our domestic animals."

"I really do want to know about that," Moses said. "It is such a problem in our village."

The pastor went on to explain that taking care of the land, the forest, and the animals requires much work. He reminded Moses that work is part of God's plan for us. He told him, "God himself worked to create everything, and he gave Adam and Eve work to do, even before they sinned. Work, you see, is good because God works. Their work produced food because the land was good. When they sinned, however, they still had to keep on working. Only now it became much more difficult, and it is still difficult. That is why many people weep when they plant the seeds they would like to eat because they are hungry."

Obeying God's Word is helpful

"Wait a minute,"Moses said. "I have a question. Adam and Eve had dominion over nature. They were to rule over the land with God. The land gave them much food. They lost that dominion when they sinned and then had to work hard to get a little food. But now you are saying that if we obey God's Word and try to follow his instructions, we can get back some of the dominion over the land he gave us. Is that right?"

Rose spoke up immediately. "Of course. Isn't that what we have done here in Katindi? We knew the wisdom of our ancestors was good, but it was not sufficient. We have learned some good things from agriculture in Europe. But we see that some things that work in Europe do not work here. We studied the Bible to find out what God says to us. We learned from our own experience that some trees protect the land and help it produce more food. We prayed and asked God to help us put all this wisdom together. Now we have more food to eat."

"That's like what we do in our school of agriculture," John said. "We study the soil to find out good ways to grow foods. What we discover is what God has created. So what we call the 'laws of agriculture' are really the laws God put into agriculture."

"I am just an old chief," Moses exclaimed, "but I am finally beginning to understand. What you have found out by observing how things grow well and what you study in agricultural school really come from God."

"That's absolutely right," the pastor affirmed. "You see now why it is important for us to follow God, both in what he tells us in the Bible and what we discover in all he has made."

Wise and unwise men

Pastor Simon said, "Before we go, let me tell you a story Jesus told that helps us understand what we have been talking about. It's in Matthew chapter 25. I'll tell it instead of reading it.

"A wealthy man was going to take a long trip. He called three of his servants into his office and gave each one a sum of money. He gave one a very large sum; to the second he gave a big amount, and to the third he gave a smaller amount. He told them to use the money wisely and make it produce more money.

"When he returned, he called them into his office and asked for an accounting. The first servant who received the very large sum gave the wealthy man twice the amount. The second one also gave back twice what he had received. The wealthy man was very pleased. He told them they had been faithful and wise in using what he had given them. He would now give them much more.

"The third man was afraid. He simply gave back the amount he had received, and the bills were wrinkled and very dirty. He had produced nothing with them.

"The wealth man was very angry. He told this servant he was lazy and wicked, and could not be trusted. He took the million shillings and divided it among the other two servants. He then told the lazy servant to leave and never come back.

"You see," Pastor Simon concluded. "God has given us a wonderful land full of many marvelous things. He requires that we use it all wisely. He will make us give an account to him of what we have done with the land. When we use it well, he will give us more good things. However, here in Katindi we had taken this good land God gave us and made it poor. So we lost the power to grow lots of food. When we repented of this, changed our thinking and our habits, and restored the land, God has blessed us. Amen!?"

The people all said "Amen." They clapped their hands and someone began singing a hymn. When all was quiet, Moses spoke up again. "Pastor Simon, is there anything we can do in Malenga to help our land become fertile now and produce more food? It will take years for trees to grow back."

Pastor Simon replied, "Moses, we have had a long day. I am tired and I think it is time we went to bed. We have seen many things and had good talks. In the morning you, John, and I can have breakfast together and then talk about what you can do to grow more food and make your people more prosperous. Now let's pray together and thank God for his Word, for the life he does give us, and for the wisdom he gives when we ask for it."

Questions for study

Read Genesis 1: 27 - 30. God made man and woman in his image and commanded them to take charge over the earth and rule over all the creatures God made. We represent God on the earth and are to rule over it with him.

1. *How can we study the land and rule over it in order to learn how to grow more food?*

2. *What wisdom did the ancestors have about growing food? Make a list of all the good traditions they had about the land and the forest.*

3. *What can we learn from God's Word about growing food? Make a list of what God tells us about growing food.*

4. *How can science help us in this?*

5. *Read Matthew 25: 14 - 30. How can we apply this parable to how we use the land God has given us?*

6. *In what ways will this help us have the abundant life Jesus wants us to have?*

Chapter 5

Why Grow Trees Around Fields?

Outside Pastor Simon's house a bright fire was burning. He, John, and Moses sat around it talking. Pastor Simon began explaining to Moses how to start planting new trees.

"In Africa we have more than one thousand different kinds of trees. Some trees are good for producing wood for building houses and for making furniture. Other trees produce fruits for us to eat. Still others are useful to make the soil in our fields and gardens stronger and able to grow more food. And some trees, like acacia, grevillea and other fast-growing trees make good firewood. Let's talk first of all about planting trees around your fields and gardens."

Moses was curious. "Why in the world should we do that?"

Border trees help us in many ways

1. A living fence

The pastor replied, "Moses, trees growing along the borders of our fields and gardens can help us in many ways. One way is to make a living fence to keep out animals. A good fence costs money, and must be replaced every few years. A living fence costs nothing except work, and it can last many years."

"I don't understand what you mean by 'living fence'," Moses said.

A living fence makes use of trees growing closely together.

"A living fence is a row of trees 1 to 3 feet (50 to 100 cms) apart. You tie palm slats, or other branches, between them to make a fence. You keep the trees cut short, not more than a meter or two high. That way they branch out and fill in the fence. This keeps animals out of your fields."

2. Erosion control

Pastor Simon continued, "A second reason to plant trees around your fields is to stop erosion. Erosion is a big problem in farming. Trees that grow around the borders of fields on slopes help stop **water erosion**. They keep the soil from washing away in the heavy rains. I have two fields on the hillside. The border trees keep the rainwater from carrying off much soil."

3. Wind erosion

John spoke up. "Some species of tall trees can be very good as border trees. Trees like ***grevillea***, and some other native trees, can grow several metres high and quite close together. Because they are tall, they protect the soil from strong winds. They keep the wind from blowing the soil away. In other words, they stop **wind erosion**.

"Both in the valleys where the land is flat, and in fields on hillsides, strong winds can blow away much soil. If you have tall trees around your fields, that will stop the wind erosion. Tall trees may not be good, however, if your fields are small. They would make much shade. But if your fields are big, a hectare or a few acres in size, tall trees are good."

He added, "In some places the wind is very strong and the fields are quite big. There it is good to plant two or even three rows of trees around the border of the fields. The trees on the outside can be shorter. The trees on the inside can grow to be several feet (or metres) high. That will really stop wind erosion."

Pastor Simon spoke up. "John, I am sure that is true. But my fields are quite small. I do not have room for two or three rows of trees around my fields. One row of trees is enough to stop wind and water erosion."

4. Firewood

The pastor then explained another reason border trees help us. "These border trees, especially the tall ones, likewise can give us firewood. Branches grow out from the trees. When they are about as thick as my arm, I cut them off, dry them, and use them for firewood."

Moses spoke up. "Pastor, this is very interesting. I had no idea trees could be so useful."

Pastor Simon smiled. "Yes, they do help us a lot. Here are more ways they can help us."

5. *Poles/or building houses*

He continued, "Tall trees can provide poles for us to build houses. Some tall trees, like ***grevillea***, grow tall and straight. You can cut the tree about a meter above the ground to get a straight pole. From the stump, the tree will continue to grow and become tall again. Of course, do not cut them all at one time. Cut only one out of every five to ten trees. That way the living fence is still there to help you."

6. *Restoring fertility to the soil*

"Moses, the other day we talked about another way they help us," the pastor continued. "Having leucaena or acacia trees around the border of the fields helps the soil produce more food."

Moses asked, "Yes, I saw that in your fields. But how can trees help us grow more food in our fields and gardens?"

"John, please answer that question for Moses," Pastor Simon said.

John began. "All trees and plants, including those we grow for food, need nitrogen in order to grow well. The air we breathe is 80% nitrogen. But most plants cannot use nitrogen from the air. They have to get it from the soil through their roots.

"One kind of tree takes nitrogen from the air and puts it into the soil. We call this kind of tree a leguminous tree. These trees produce long beans that split on a single seam. Leucaena and acacia trees, and some other types as well, are like this. Uncle Simon, I think you have leucaena trees around your fields."

"That's right," the pastor replied. "Moses, I have planted both leucaena and acacia trees all around the borders of my fields and gardens. Their roots add nitrogen to the soil and that helps crops to grow better. Tomorrow we will go see my fields and the trees that make the soil better."

7. Mulch for the crops

"The branches of both leucaena and acacia trees make good mulch," John said. "However, acacia branches grow long thorns, so you can use only the new young branches before they grow sharp thorns. The branches of grevillea trees also make good mulch. You cut some of the new branches of the border trees and lay them on the ground between the plants of maize, beans, millet, or other crops. The leaves decompose and make compost to help the crops grow better. This mulch also keeps weeds from growing. When it rains, the rain goes down into the soil. The mulch keeps the hot sun from drying up the soil quickly. So the soil stays moist longer, and that helps the crops grow better."

8. Food for animals

Robert spoke up. "Their branches also make good food for the animals. I feed them to my goats and sheep. This helps them grow well."

"You see, Moses. Growing trees around the borders of your fields can help you in many ways," the pastor said.

9. Alley cropping

Then he added, "I also do something else very helpful with these leguminous trees. In my fields on the hillside, I have made parallel rows of Acacia angustissima five metres apart across my fields, horizontally along the slope, so I can plant crops between them. We call this 'alley cropping.' It really helps me grow more food in these fields every year. The roots make the soil rich by putting nitrogen into the soil to help the crops grow well. The branches cut from them cover the soil and protect the growing crops."

Growing trees for borders and for fields

"How did you get those trees to grow in your fields?" Moses asked.

Before Pastor Simon had time to answer this question, Pastor Henry from Kaledi and a group of men and women from his church walked into Pastor Simon's yard. Pastor Henry greeted Pastor Simon and introduced the group of people with him. He explained, "We understand you are helping people learn how to grow more food. Our people have difficulty growing enough food. So we have come to talk with you."

"Wonderful," Pastor Simon replied. He and John quickly brought more chairs and put them in a circle. "Please sit down and join our conversation. This is Chief Moses from Malenga, and we are discussing how planting trees around your fields helps you grow more food."

Pastor Simon and John went over the ideas they had been discussing. They explained to the group from Kaledi how planting trees around fields and gardens can help grow more food.

Pastor Henry then said, "We don't know how to plant trees. Can you help us with that?"

"Of course," Pastor Simon replied. "You have joined our group at the right time. Moses has just asked us about that. We can now explain to you all how to plant new trees."

At that moment Virginia, Pastor Simon's wife, came outside with a big pot of tea and many cups. She began pouring a cup of tea for each person in the group. Her daughter Denise followed her with a plate full of biscuits. As the group began drinking tea and eating the biscuits, they thanked Mrs. Pastor Simon and her daughter. They also thanked Pastor Simon for allowing them to come into this discussion.

Questions for study

1. *What are good trees to plant around fields and gardens?*

2. *How do these trees help the fields produce more food?*

3. *The Bible talks about the 'tree of life.' We find this in Genesis chapter 2 and Revelation chapter 22. Why do you think God chose the tree as a symbol of life?*

Chapter 6

Growing New Trees in a Nursery

When they had finished drinking their tea, Pastor Simon began explaining to them how he plants trees around the borders of his fields. "I grow them from the seeds of those leucaena or acacia trees," he said. "There are two ways to plant these seeds. One way is to plant them directly in the soil along the borders of the fields. First, I use my hoe to turn the soil well and make it loose so the seeds can grow easily. Then I add compost or manure to make the soil rich. I do this at the beginning of the rains. That way I need to put water on the little trees only when we have more than three days without rain."

He continued, "The other way is to plant them in a nursery. When they are growing well I transplant them to where I want them to grow."

John spoke up. "We do a lot of that where I go to school. We have a big nursery there for growing new trees."

"What is a nursery?" Moses asked.

Starting a nursery

"It is a special place for growing new plants or trees," John replied. "It takes a lot of work and care. But it is an excellent way for starting many new trees, trees that can really help your garden and fields. It is something you can do yourself. Or the whole community can do it together as a community project."

"How do we start a nursery?" Moses asked.

John replied, "First of all, find a place outdoors near your house where there is shade. The young trees need protection from the hot sun. Probably a space fifteen feet (five metres) by eight feet (2 1/2 metres) will do if you do this yourself. It should be bigger if the community wants to grow many trees. If you do not have an area that has good shade most of the day, then you need to build a shelter of sticks over the nursery and cover it with branches of trees. Some sun must get through, but not much. If you have some money, buy a heavy net and put it over it to protect it from the hard rain and from birds. You will also need water so you can put water on the new trees every day."

Planting new trees in plastic bags

"How do I start growing these new trees?" Moses asked.

John answered, "One way to do it is to plant the seeds in plastic bags. You need many plastic bags, because you grow each new tree in one plastic bag. The bag needs to be big enough to hold about a quart (litre) of soil. Fill each bag with good soil, soil that has some compost in it and also some sand to make it loose.

"Find good soil somewhere, perhaps down in a valley by a stream. Then find manure from goats, chickens, or cows. Take four shovelfuls of good soil, one shovelful of manure, and one shovelful of sand. Mix them well together, and then begin filling the bags. Put a little into a bag and press it down gently. Then put more in and press it down again. Do this 3 or 4 times until the bag is almost full.

"Make small holes in each bag on the bottom and the low part of the sides. These will let water drain out. You want to keep the soil moist but not too wet. The bags are now ready for you to plant tree seeds."

Planting new trees in seedbeds

Pastor Simon spoke up. "John, plastic bags cost money and are sometimes hard to find. You can have a tree nursery simply by planting tree seeds in seedbeds. You then transplant them when they grow into little trees. The seedbeds should be raised about 20 cm above the ground. The soil in them needs to be loose and have lots of good compost or a bit of manure in it to make the seeds grow well."

"That sounds good to me," Moses said. "I don't have money to buy bags. I could take a small spot near my house and make a nursery right on the ground."

Nursery with small trees growing in bags and in seedbeds.

"Fine," the pastor replied. "Make the soil there as good as you can with compost and manure. However, be sure chickens, goats, and other animals cannot get into it. They can destroy a nursery in just a few minutes. So a strong fence is essential."

Bags or seedbeds?

John looked worried. "Uncle Simon," he said. "In spite of the difficulty, planting new trees in bags is better. Here are some reasons why:

1. It takes much water to water new trees in seedbeds, but only a little water when they grow in bags.

2. When the new trees have sprouted but are still small, they will need more sun. You can easily move the bags to a place that has more sun. You can't move little trees that are growing in a seedbed. Of course, in that case, you can remove some branches from the shelter to let more sun come through.

3. It is easy to transplant seeds from bags into the place you want the trees to grow. All you need to do is carry the bag to the place you want to plant the tree. If the place is far away, carrying a bag is easy. Carrying a new tree dug up from the seedbed can be difficult. Much damage to the roots can occur. But when the little tree is still in the bag until it gets to the hole, no damage occurs."

John concluded, "So I would suggest you do all you can to find plastic bags. Of course, if you can't find them, or do not have money for them, then you can make seedbeds."

"Where do I get tree seeds?" Moses asked.

Planting tree seeds

John said, "Seeds from acacia and leucaena trees are easy to find. I am sure Pastor Simon can give you some because each tree produces many seeds. Look on healthy trees for the best seeds, ones that are big and look good."

"How do I plant them?" Moses asked.

"See how big each seed is. Push a finger into the soil in the seedbed or in the bag to make a small hole. It should be as deep as two times the width of the seed. Put 3 to 5 seeds in each hole. This is to be sure that at least one will grow. If three or four grow, pull out all but the best looking little tree. You want only one new tree from each hole or in each bag."

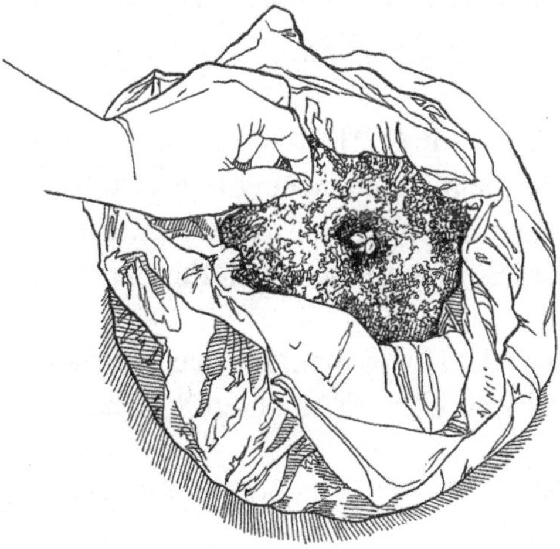

Planting seeds in a bag.

Pastor Simon added, "You can also use small pots of clay or plastic instead of plastic bags. The pots should each have a hole in the bottom to let the extra water drain out. The top of the pot should be bigger than the bottom to make it easy to remove the young tree when you want to transplant it."

John continued, "Water the seedbed or these bags or pots every day in the late afternoon. If the seeds are in bags, put in water until it begins to run out the little holes near the bottom of the bag. That will be enough to keep the soil moist."

"Water is such a problem," Moses said. "The women have to bring water every day for a long distance and we never have enough.

John laughed. "Yes, I know what you mean. You know, however, that every day you throw out a lot of water – from cooking, washing your hands, or taking a bath. Don't waste this water. You can use it for your garden and your nursery."

"Even when it has soap in it?" Moses asked.

"Yes," John replied. "However, when the water has soap in it, don't pour it on the plants or young trees. Pour it carefully on the soil at the base of the plants or trees. That way the water gets into the soil, but the soap will not harm the plants or trees."

He continued, "If you plant the seeds directly into the seedbed in the nursery, plant them about 8 inches (20 cm) apart."

Moses asked, "When is a good time to start this nursery?"

Pastor Simon replied, "Start it in the beginning of the dry season and when you can find good seeds. You want to have the young trees ready to transplant when the rains start. That way you will not have to water the transplanted trees every day. New trees need four to six months in the nursery. The stems should be about as thick as a pencil before you transplant them. Leucaena and acacia grow more quickly. But even they need to be as thick as a pencil before you transplant them, or even thicker."

Transplanting trees from the nursery

Moses asked, "Where should I plant these trees when they get big?"

Pastor Simon replied, "Plant leucaena or acacia trees around the edges of your gardens and fields. They will soon put nitrogen into the soil. That helps the crops to grow better. I will show you tomorrow those trees that are growing around our fields here."

"How far apart should I plant them?"

"That depends," the pastor replied. "If you plant them close together, say every three feet (1 meter) or less, they can make a living fence around your field that will keep out the cattle and goats. Tie palm slats, sticks, or bamboo strips between them to make a tight fence. If you already have a fence around your field, you can plant them every six to ten feet (two or three metres)."

Moses said, "I think I understand this. Now tell me how the young trees growing in the nursery should be planted where they can grow into big trees."

John said, "We call that 'transplanting.' The young tree should be at least 8 inches (20 cm) tall. The stem should be as thick as a pencil, or more. If the little tree is growing in a big bag, the tree can be even a foot and a half (50 cm) tall.

"First, dig holes where you want to plant the young trees. Each hole should be quite big, up to 2 feet (60 cm) wide and 2 feet deep. That way the roots of the little tree can grow easily. Put the topsoil from the hole on one side. Then put the soil from deeper down on the other side.

"Early in the morning before transplanting the trees, put water on each young tree so the soil will be moist. If the little trees are in bags, take all the trees you want to plant at one time to the place where you will plant them. Put one tree in its bag by each hole."

"How do I get each tree into the ground?" Moses asked.

Protect the roots of the little trees!

John answered, "It is important to keep the soil around the roots of the young tree. That prevents damage to the delicate roots the little tree needs in order to grow well. If the little trees are in bags, it is easy. Cut or tear the plastic bag carefully down each side. Take it off the ball of soil around the roots. Try to keep the soil from falling apart. Put your hand under the ball of soil with the young tree and put it carefully down into the hole. Try to make sure that all the roots are planted downward into the soil. The roots may be coiled at the bottom, or growing sideways. If so, gently try to pull some of them down. The little tree will grow best if all the roots are going down.

"If the young tree is growing in a pot, turn the pot carefully upside down. Gently pull down on the young tree and it will come out with the soil around the roots."

Moses said, "I will probably make my nursery right in the ground, like Pastor Simon said. I don't have money to buy bags. How do I get those little trees to the place where I want to plant them?"

Pastor Simon answered. "This needs to be done very carefully. Protect the roots well. First I water each tree well in the seedbed. I take a shovel and cut down into the soil around each little tree. If the little trees are 8 inches (20 cm) apart, cut the soil around each one 4 inches (10 cm) from the stem. I go down as deep as I need to – more than a foot (30 or 40 cm) - to be sure to get all the roots. I lift each tree out carefully to keep the soil around the roots. That way they will not be damaged. I carry each little tree with its ball of soil in the palm of my hand to the hole where I want to plant it. I put it directly in the hole. I do this transplanting early in the morning, or late in the afternoon. This reduces stress on the little tree."

Lifting a little tree out of its hole in the nursery.

John added, "Make sure the roots have plenty of room to spread down and out. If the ball of soil seems quite hard, loosen it a bit with your fingers. Then put the topsoil from the hole gently around the ball of soil with the roots of the young tree. Finish filling the hole with the soil from deep in the hole.

Putting a new little tree into a hole in the ground.

"Do not build up soil around the stem, or trunk, of the young tree. Fill the hole only up to the level of the soil from the bag in which the tree has been growing. Then gently press the soil down around the young tree with your hands. That prevents pockets of air staying in the soil around the roots. Air pockets make the roots dry up. Finally, put more water around it. The soil around the tree should be a little below the ground around it. That way rain water will collect there and go down to the roots of the little tree."

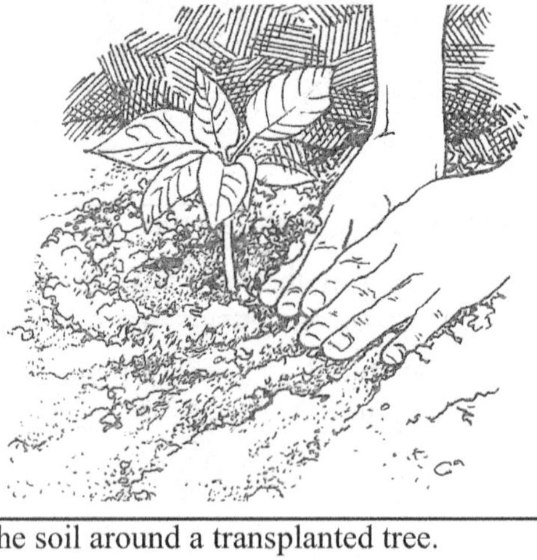

Pressing down the soil around a transplanted tree.

Water and protection

Pastor Simon added, “One other thing is important. You need to work together to make sure that cows, goats, or other animals will not go where you have planted the new trees. If they do go there, they can destroy them all quickly.”

Fence protecting a newly planted tree from animals.

“Do I have to keep watering the newly planted trees?” Moses asked.

“That depends on the rain,” John answered. “It is a good idea to do the transplanting at the beginning of the rainy season. If a good rain comes once or twice a week, you will not need to put water on the trees yourself. But if the rain does not come, you should put up to a bucketful of water on each young tree twice a week.”

"Whew, that is a lot of work," Moses exclaimed.

"Of course it is much work," the pastor replied. "But it is good work because it will increase the amount of food you can grow. You say you get little food from you fields now. But in two or three years you will get much more food from your fields. That is because of the nitrogen the acacia or leucaena trees put into the soil. I will show you tomorrow how this has helped my fields produce more food."

"That is why it is better to transplant the young trees during the rainy season," John said. "You can plant the seeds in the tree nursery any time, because there it is easy to put water on them. And the new young trees do not take much water."

Pastor Simon added, "It's important to keep weeds from growing around the little trees until they get quite big. Weed around the trees for a distance of 2 to 3 feet (50 to 100 cm)."

"Can you grow fruit trees in the same way?" Moses asked.

Pastor Simon smiled and said, "That is another big subject and an important one. Are you tired or can we talk about it now?"

"Sure, let's talk about it now," Moses answered. "I want to learn all I can from you and John."

Questions for study

1. *What must we have and do to start a tree nursery?*

2. *How can a church, a school, or a whole community start a nursery?*

3. *Where can we find the seeds of the trees we want to plant?*

4. *What are all the things we must we do to help the seeds grow into trees?*

5. *What are the rules to follow when we transplant a tree to the place where we want it to grow?*

6. *It we transplant a little tree and it dies, what could be the reasons why it died?*

Chapter 7

Growing Fruit Trees

Planting citrus or guava trees

John began talking with Moses about growing fruit trees. "You can take the seeds from fruits like citrus or guavas and plant them in a nursery. Place the seeds on a mat or board in the shade where they can dry out completely. Do not cover them but let the air get to them."

Pastor Simon added. "You can collect the seeds any time of the year. When they are dry, store them in a closed jar or bottle to keep them dry until you plant them."

Collecting seeds from citrus fruit and soaking them in water.

John added, "Fruit trees sprout and grow slowly. So plant the seeds in bags or in a seedbed 3 to 6 months before the rains come back. The young trees will be ready to transplant when the rains start. You can help the citrus and guava seeds sprout better by soaking them in water for two days before planting them in bags or directly in the ground. Or you heat a can of water so that it almost boils. Take it off the fire and drop the seeds into the hot water. Leave them there, and let the water cool overnight. That will make the shell of the seeds soft."

"Can I plant seeds directly in the ground where I want the trees to grow?" Moses asked.

"Yes, you can do that,"Pastor Simon replied. "You should do that at the beginning of the rainy season. Plant two to five seeds in each spot. If more than one sprouts, pull out all but one. You must water them regularly if there is no rain for three days or more. Pull out the weeds around the young sprouts so the weeds do not crowd them out. Be sure to protect the young trees from animals."

Planting citrus trees by cuttings

Pastor Simon now spoke up. "Moses, you can also plant citrus trees by taking cuttings from a good tree and planting them directly into the soil where you want new trees to grow."

"How do you do that?" Moses asked.

Pastor Simon replied, "First, till the soil where you want the new fruit trees to grow. Make the soil loose so the cuttings can make roots easily.

"Find healthy branches on a good citrus tree, branches at least one year old. They should be from one to three cms thick, like your thumb or middle finger. Cut each branch off with a sharp axe or machete. Each one should be about up to a foot long (1/2 a meter). Cut off the far end where it gets thin. Place them on a flat board. Cut the tree end (bottom end) with a sharp machete in a diagonal, or slant, way to make a very clean cut. Strip the leaves off each cutting except for one or two leaves at the upper or outer end of the cutting. That way you can tell which end to push into the ground. You push in the slanted end. The end with the leaf or leaves will stick up out of the ground.

"Plant these cuttings in the loose soil. Push them into the soil for at least half their length. Make real sure you put the end you cut from the tree down into the soil. That is very important, for the cutting cannot grow if you put the top end of the cutting in the soil. Put some water on each one. Plant the cuttings at the beginning of the rainy season. Be sure to water them every day that it does not rain."

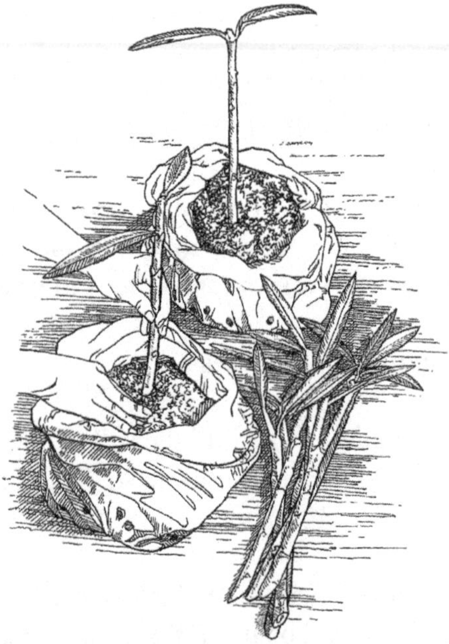

Planting cuttings in plastic bags.

John spoke up. "If you have a few plastic bags, you can plant the cutting in them. That way you can plant them any time. When they are ready to transplant, you can take the bags right to the spot where you want the tree to grow.

"New buds and leaves will start growing soon. But it will take two or three months for roots to grow in the soil. Only then will the cutting begin to grow into a new tree. Some cuttings will dies. Do it well and perhaps half of them will become a new tree."

Grafting citrus trees

"My problem is this," Moses said. "We have orange and lemon trees in Malenga, but they produce sour fruit. We don't want to grow more of those trees."

"There are two answers to your problem," John said. "The first is to come to the agricultural school and get seeds from sweet orange, grapefruit, and lemon trees. However, those seeds may not grow well in your village.

"The other answer is a better one, but it takes some skill. That is to graft onto your sour orange and lemon trees branches from sweet orange and lemon trees. Your sour trees grow well in Malenga and they provide good roots for the trees. You then graft branches of sweet orange or lemon trees onto those old sour trees. These branches will grow and produce sweet fruit."

"How in the world do you do that?" Moses asked.

"It takes skill to do it right," John said. "If you like, send one or two of your men to the agricultural school and we will teach them how to do it. They can then take small branches from sweet trees to graft onto the sour trees you already have."

"That sounds like a good idea to me," Moses said. "We do want sweet oranges, lemons and even grapefruit. We'll talk it over and see who we can send to your school."

Planting mango and avocado trees

John continued. "Now let's talk about mango and avocado trees. We can grow them directly from the pits. Take healthy-looking ripe mangoes or avocados from a healthy tree. Remove all of the fruit from them. Let them dry a few days. Then crack them open or break the bigger pits from mangoes or avocados before planting them. That helps them sprout faster."

Pastor Simon added, "Plant them right away. After two weeks they are no longer alive. Plant them when the fruit is ripe. Take the husk off the outside of the mango pit. Lay the pits in a shallow hole, so that

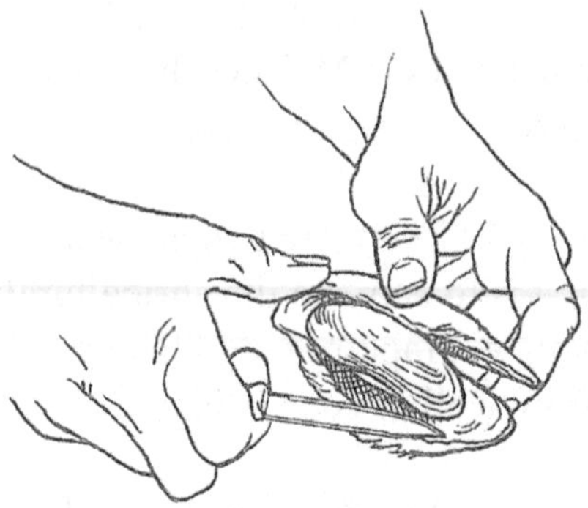

Opening a pit with a knife.

Avocado pit planting and growth

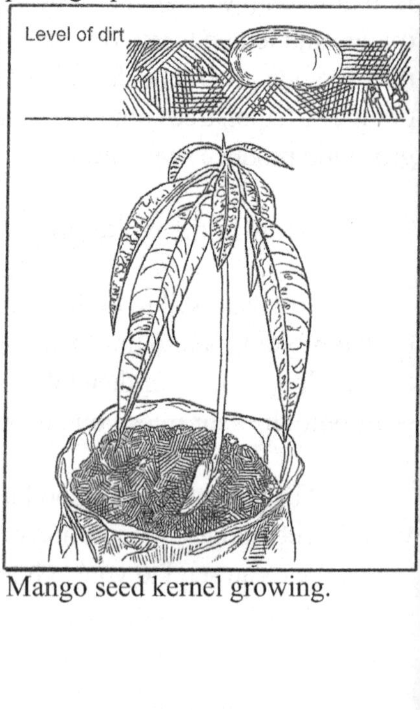

Mango seed kernel growing.

part of the pit is not covered. The mango pit should lie with the big curve up and above the soil. The avocado pit should have the smaller end sticking up above the soil."

John said, "Keep the soil around them wet, especially after you see the sprouts coming up. You will need to water the little trees regularly through the next dry season until the roots have gone deep enough to get water from far down in the soil."

The pastor continued, "Weed around these little trees so they are not choked out by weeds. Build a short fence around each one of them to protect them from animals. Goats can destroy the young trees quickly."

Then he added, "Moses, we have a proverb like this." From a mango seed we get one tree. From one tree, we can get thousands of new trees.' So remember that. If you have one good mango tree, you can get all the new trees you want."

Planting banana trees

John asked Moses, "Do you have lots of banana and papaya trees?"

"A few of our people do, but many don't," he replied.

"Did the people who have them plant them?" John asked.

"Oh no," Moses answered. "They just happen to grow near their houses."

"Well, let's talk about how you can plant banana and papaya trees so that you will have bananas and papaya almost all year. We will start with banana trees.

"Banana trees do not start from seeds. New trees start from little shoots that we call 'corms.' They grow up next to a mature tree. You will see a new corn with small green leaves sprouting up near the big tree. We call this corm a daughter tree.

"To start a group of banana trees at your house, find a strong looking corm that is growing up alongside a healthy mature tree. The corm should have long leaves, perhaps like long machetes, or perhaps quite broad. Dig this up and shake or brush all of the soil from it.

Banana trees with daughter corms ready to transplant to a new location.

"A real problem in growing banana trees is microscopic worms in the soil called nematodes. You can't see them with your eyes. They are too small. You can do one of two things to get rid of the worms from the corm. Take a machete and cut off all the little roots. Cut right down into the flesh of the bulb of the corm. This will get rid of most of those tiny worms. Or you can put the corm into a pail of hot water, not boiling, but steaming. Leave it there for twenty minutes or so. This will kill any tiny microscopic corms that can damage the young tree.

"Dig a hole about the length of your arm, and about the same width across. Plant the corm in that hole. Pack the soil firmly around it, and then pour in a bucket of water. You must pour in water every day if it does not rain. You can use bath water or kitchen water for this. You can plant two more healthy corms nearby so as to have three bananas trees growing in a group."

Pastor Simon continued. "Put a lot of mulch around the base of each corm – dead leaves or grass, garbage from the kitchen, and a few droppings from goats or cows. If you put a lot of mulch around it, it will grow bigger and faster. Keep weeds from growing close to the young trees because they will keep them from growing well."

John continued, "A new tree will give a bunch of bananas as soon as 9 months, although it may take up to 18 months. The tree will bear only one bunch, and then it will die. By that time, the corm, or daughter tree, that is growing near it will replace it. You will then see from three to six other smaller corms sprouting up near the mother and daughter tree. They are the 'grand-daughter 'trees. You must pull up all but two or three of them. Those corms will then replace the daughter tree after it has had a bunch of bananas."

Pastor Simon explained, "Ideally, you should only have three generations of tree in one place – up to three mature trees giving a bunch of bananas, 3 healthy daughter trees to replace the mother trees, and two or three small grand-daughter trees around each one that will replace the daughter trees. If you let more than 3 corms grow for each mature tree, they will take food away from the mature tree. None of them will be healthy. Only a few of them will produce bananas."

John went on. "Keep on putting mulch around the base of the banana trees. This will help them grow strong and produce big bunches of bananas. When a new bunch has begun to grow, put a long big stick in the ground under it so that the top of the stick will help hold up the regime. The bunch will soon become very heavy and may make the tree fall down. So this stick will help hold up both the bunch of bananas and the tree."

Pastor Simon said, "If you want more bananas, and you probably do, plant a healthy-looking 'grand-daughter' corm in another spot. This will start another group of bananas growing in the same way. If you have several groups of banana trees growing, you will have bananas almost all year round."

John said, "I mentioned that tiny worms live in the soil that can harm banana trees. Because of this, every five years, start anew group of banana trees growing in a different place. Cut down the old groups, and use that land for growing vegetables."

The pastor added, "If you want to grow cooking bananas, called plantains, you grow them in just the same way."

Growing papaya trees

"What about papayas?" Moses asked.

"Papaya trees grow from seeds," John said. "Choose a nice looking papaya from a healthy tree of a good variety. Open it and put the seeds in a pan of water. Rub them together vigorously to take off the outer covering of gelatin. Then let them dry in the shade until they are completely dry. You can keep them in a closed bottle or can until you are ready to plant them. But plant them as soon as possible.

"You can plant the seeds in plastic bags, or in a seedbed. The best way is to plant them in the places where you want the trees to grow. If you plant them in the nursery, do it two months before the rains start. Put 3 or 4 seeds in each bag. When they sprout, pull out all but one, the best looking one. If you plant them in a seedbed, plant them at least ten cm apart. When they are 20 cm tall, or the length of one half your forearm, they are ready to plant.

"When it is time to transplant the young trees, choose a good spot where the soil is somewhat humid or moist. It should have good drainage so water does not stay in the soil a long time because too much water can quickly kill the seedlings. Plant the young trees about two metres apart. Put mulch around them. Water them two or three times a week if it does not rain."

Pastor Simon added, "Be very sure to protect the roots of the young papaya trees. They are fragile. Dig each one up carefully and take it immediately to the hole where you will plant it. Keep the soil tight around the roots. This is why it is usually better to plant the seeds directly where you want the new trees to grow. Then it is not necessary to transplant them."

He continued. "A papaya tree is either male or female. You need only one male tree for every group of female trees."

"You mean they practice polygamy?" Moses asked.

"Well, sort of," Pastor Simon replied, and the whole group laughed. "But I don't think God minds. He made them that way." Then he smiled and said, "Remember, God made us different from papaya trees."

Female papaya flowers.

Male papaya flowers.

Breaking off the top of a young papaya tree to keep it from growing tall.

John said, "Keep the young trees from growing very tall. The fruit is hard to pick from a very tall tree. Pinch off the center of the top of each tree when it is about as tall as you are. That way, the tree will stop growing tall and will bear fruit where you can easily reach it.

"A tree will produce fruit for two to six years. Every few years, start growing trees in a new place. Cut down the old ones and use that land for other crops. This will protect the papaya trees from many diseases. If you see a disease starting on the leaves or the trunk of a tree, cut the tree down immediately with its fruit and burn them up. You will lose a few papaya trees that way. But if you do not cut down the trees with disease, the disease will go to all the trees. Then you will lose them all."

Moses said he was very grateful for all this information. He said the farms of the people in his community were small and they have much difficulty in growing enough food.

Food for all the family

"Actually," Pastor Simon said. "It is possible for a farmer with just one hectare of land (two and a half acres) to grow enough food for himself, his wife and his children."

"How can that be?" Moses asked.

"Planting leucaena, acacia trees, grevillea and other trees around the border makes the soil near the trees stronger, provides firewood, and keeps out animals. The farmer should use two thirds of the land to grow grains like maize, millet, or sorghum and maybe cassava. A few big acacia trees in the middle will give some shade and make the soil stronger. Tomorrow I will show you what we do here, and you can see if that will work well in Malenga.

"The other third of the farm would be for the house, a vegetable garden, fruit trees, a woodlot of trees for firewood, and maybe space for some chickens or small animals. Growing a good amount of grains, plenty of vegetables and having lots of fruit will provide an excellent diet for the whole family."

John added that his agricultural school has many small books about how to grow different kinds of fruit trees, trees for firewood, and forest trees from seeds. These books are important to show how to plant the different kinds of trees.

Moses thanked the pastor and John for all of this information. He said he really wanted to see how this was working in the pastor's fields and would be happy to go there in the morning. He then said to the pastor, "You said we should also plant trees for firewood. Can you tell me about that?"

John replied, "It is good you asked that question. Next month, in the district center, we will have a meeting of community leaders and teachers from the agricultural school to talk about this. It is a very important issue. We all need wood for fuel, and that is why we cut down trees and destroy the forest. We will be talking about how to get enough wood for fuel without destroying the forest. In fact, each farmer should grow trees to give himself sufficient wood. Some farmers can actually grow trees to sell for firewood and charcoal. So please come to that meeting."

The pastor smiled. "John, that sounds great. Can Moses and I come, and also invite other pastors and church leaders to that meeting?"

"Of course," John replied. "The church can do much to help with this issue."

Pastor Simon smiled and said, "I think we are all getting a bit sleepy. Let's thank God for the wonderful discussion we have had here. Then we can all have a good sleep. In the morning we will look at our fields here in Katindi."

Moses and John agreed. The pastor thanked God for the good discussion. He asked Him to continue to show them how to restore the soil and the forest and how to grow more food.

Questions for study

1. *What are the steps to follow in growing new citrus or guava trees from seeds?*

2. *How can you grow them from cuttings?*

3. *Why must new little trees growing in a nursery be transplanted very carefully?*

4. *What must you do to start new mango or avocado trees?*

5. *What is important to remember about growing banana trees? And about growing papaya trees?*

6. *How can a man with only a small amount of land grow enough food for himself and his family?*

7. *Jesus said, "Do not worry about your life, what you will eat or drink." Matthew 6: 25. Does that mean we should not work hard to grow crops and trees?*

Chapter 8

Rotating Fields to Grow More Food

The next morning Pastor Simon took Moses and John out to see the fields of the people of Katindi. They went first to Pastor Simon's fields. They talked together as they went along.

The old method of planting fields

Pastor Simon said, "In the old days, we changed our fields very often. We could grow crops in a field for two harvests. Then we had to let the land rest for five harvests. So every two years we had to find new fields."

Moses said, "I remember my grandfather telling me that fields had to rest for several years before being used again. But I never understood why."

John spoke up. "I explained to you yesterday how food for growing crops gets into the soil. It comes from the compost made of leaves, grasses and weeds that fall on the ground. It also comes from the minerals that roots bring up from deep down in the soil. In the old way of farming, the land must rest for four or five years for compost to develop and roots of the trees and brush to bring minerals up from deep in the ground. The soil will be fertile again."

Moses said, "Pastor, tell me what you do now to make your land more fertile."

The new method of planting fields

Pastor Simon answered, "I have divided my land into four fields. You can see two of them here. The other two are over there on the side of the hill. In this field I am growing millet and beans. Some of my neighbors prefer sorghum, but we like millet. In the second field, I am growing maize, with groundnuts growing between the stalks of maize. In the third field on the hill you see cassava growing. I planted it this year along with squash."

Moses said, "I see only cassava in that other field on the hill."

Pastor Simon said, "Yes, that is right. That fourth field is resting this year. But the cassava I planted in it last year is still there. We can harvest it whenever we need it.

"Next year I will plant millet and beans in that fourth field. We will have eaten all of that cassava. I will plant maize and groundnuts in the first field where the millet and beans are growing now. I will plant cassava and squash in the second field where there is now maize and groundnuts. That third field of cassava will rest next year."

John said, "Are you sure you will have enough cassava to eat? It grows very slowly."

Pastor Simon replied, "Oh yes. While the cassava is growing in the third field, we can harvest the cassava in the field that is resting. Next year, the cassava we plant this year will be ready to eat. So there is enough cassava in the two fields for two years. While that field on the hill rests next year, we will still harvest the cassava from it. But I will not plant anything new in it, for it must lie fallow."

Pastor Simon pulled out his Bible and said, "God himself tells us that the land must rest. Here is what he says:

> *When you enter the land I am going to give to you, the land itself must observe a Sabbath to the Lord. For six years sow your fields and for six years prune your vineyards and gather their crops. But in the seventh year the land is to have a Sabbath of rest, a Sabbath to the Lord. Do not sow your field or prune you vineyards ....The land is to have a year of rest. Leviticus 25: 2-5 NIV*

You see, God knows the land needs time to recover its fertility. So that is why he gave us these instructions."

The importance of rotating crops

John remarked, "Our teachers call this *crop rotation*."

Pastor Simon asked, "Do your teachers explain why that is important?"

"Oh yes," John said. "One crop takes certain minerals from the soil. Another crop takes other minerals. If you plant the same crop each year in a field, it will take out all of certain minerals. Then the soil will no longer be fertile.

"They also teach us that certain crops like groundnuts, beans, and soya make the soil more fertile. They take nitrogen from the air and put it into the soil. All plants need nitrogen to grow and produce food. So these crops help the soil become more fertile and produce more food. Have you tried planting soya, Uncle Simon? It has much protein."

"No, I haven't," replied Pastor Simon. "They grow soya in

villages the other side of the river. Soya grows well in the soil there, but it does not grow well here. However, groundnuts also have much protein and they grow well here."

He said to Moses, "Three things are important here. The first is that I mix crops in the same field. Grains, like maize and millet, take a lot out of the soil. On the other hand, legumes like beans and groundnuts put nitrogen back into the soil. We call them nitrogen fixing crops. Growing a grain crop and a legume crop together is good.

"Secondly, I change the crops in each field every year. What one crop takes from the soil this year will come back next year when that crop is no longer there. This way the soil can keep its fertility.

"Finally, by doing this, I can use these fields every year. It is no longer necessary to let fields rest a long time. One year of rest every four years is enough. This is why I can plant trees on the borders of my fields for I have the same fields each year."

How trees around the borders help grow more food

Pastor Simon pointed to the rows of trees all around the border of his four fields. "Moses, do you see the trees growing on the edges of each field?"

"Yes, I do," Moses said and smiled. "Now I understand how they help the fields produce more food."

"I planted those trees around the outside of each field," the pastor said. "My neighbors have also done the same thing around their land. Some of the trees are acacia; others are leucaena. You can see that the crops are good near those trees around the edges. They use the nitrogen the roots of those trees put into the soil."

John added, "Other kinds of trees also fix nitrogen. Acacia and leucaena are the ones we find best here."

Moses asked, "Do those rows of trees keep the wind from blowing the soil away?"

Pastor Simon answered, "Oh yes. When they are ten feet (three meters) high, the wind blows away little soil. Dust storms do not happen here. We call those rows of trees a wind-break."

Using tree branches for firewood

Moses asked another question. "Do you get firewood from these trees?"

Pastor Simon answered, "Oh yes. When they die we use them for firewood. Sometime I cut a leucaena tree off three feet above the ground. I use the top of the tree for firewood. Then the stump will grow new shoots and the tree will keep growing."

"We also cut one or two branches each year from our bigger trees. We even do it from mango trees like the one you see growing over there. I cut those branches into short lengths. I pile them to dry under a shelter in back of my house. When those branches dry up, we use them for firewood."

Moses asked, "You mean you do not wait for the big tree to die before you use it for firewood?"

"That's right. You can cut a branch or two each year from trees like acacia or mango trees. From the stump of the branch one or more new branches will grow. The tree remains alive. That way a living tree can give you firewood for many years. By cutting one or two branches from each tree every year, you can have firewood all year long without cutting down whole trees."

Cutting leucaena trees. Branches will soon grow out from the trunks.

Moses spoke again. "I think I am beginning to understand. You said that in the old days you had to change your fields every two years and then let them rest for five years. Now you don't have to do that any more. You can keep on using these fields year after year."

"Exactly," said Pastor Simon. "The trees protect this land. The leucaena and acacia trees add nitrogen to the soil. The groundnuts and beans also add nitrogen, and I plant them in different fields every year. The dead plants and weeds and the branches of the leucaena trees add compost to the soil and protect it from the sun and hard rain. So the soil stays strong and does not become infertile. That is why I can keep using these four fields every year. Let's go look at those two fields on the hillside. I want to show you about 'alley cropping.'"

The men walked up the hillside and into one of Pastor Simon's sloping fields. The pastor pointed to long rows of leucaena trees growing close together across the field. The rows were about 15 feet (5 meters) apart, and cassava and squash were growing between them.

"Moses, this is what we call alley cropping. I have planted those leucaena trees in rows, with an alley for crops in between each row. I cut the leucaena trees regularly so they are only about three feet (a meter) tall. I also cut the branches frequently to put on the soil for mulch between the cassava and squash plants. With this mulch, and the nitrogen the leucaena tree roots put into the soil, my crops grow very well."

Alley cropping: rows of short leucaena trees with crops between them.

Growing trees to make money

The men walked into the fields of Pastor Simon's neighbors. They came up a little hill where they saw different kinds of trees growing quite close together. No crops were growing between the trees.

Pastor Simon explained. "This is Arthur's land. He has other fields for growing crops. Here he is letting these trees grow for other purposes."

He pointed to several tall trees. "Those trees he will cut down for lumber. He will make much money from the boards coming from those trees."

He showed them other trees. "He sells the branches of these trees to a man in another village, who makes charcoal out of them. The trees keep on growing and producing more branches. Arthur makes money from these trees as well. He cuts these leucaena trees regularly to sell as firewood."

Moses pointed to some milkweed trees growing here and there. He asked, "Why is he letting those trees grow? They are not good for anything."

Pastor Simon replied, "Oh yes they are. People buy the branches and the trunks of those trees and pay good money for them. The wood is very hard, and the termites will not eat it. The branches and trunks make excellent material for building a roof, a granary, or even the floor of a latrine."

When they went back to Pastor Simon's house, they sat and talked. Moses said, "I did not know that trees could help us in so many ways. I am happy to see all of this."

John said, "I have learned much myself. I can see now that the European method of cutting down all the trees is not a good method for us here. That harms the soil. It also destroys trees that are valuable for firewood, for building materials, for compost, and for stopping the strong winds. The European method of plowing is also not good. The 'no plowing method' seems much better."

"Yes," said Pastor Simon. "Trees protect the soil from the sun and hard rain. They help hold the rain in the soil. The soil does not dry out so fast, and this helps the crops grow better. Keeping the ground always covered and never plowing it is of great help as well."

Who owns the land?

Moses thought for a few moments, and then he said, "I see another problem. I am the chief in my community. Each year I tell each family what land they can use for fields. No one uses the same field every year. But here you have the same fields every year. It's like they now belong to you. How can that be?"

"Moses, that is a big question. It's about our understanding of the land. In Malenga, you follow the traditional understanding about land. The community owns the land, and you, as chief, tell the people which fields they can use."

"Here in Kalindi, however, we see that life has changed in many ways. We talked about this. The old ways no longer help us grow enough food for our big families. If we want to grow more food, we must change how we use the land. We discussed this for a long time. We believed that a family who has the same land every year will work hard to improve it. They will try to make the soil fertile so it can grow more food. So we now let families who work hard and improve their land keep that land as long as they continue doing so."

"Perhaps your way is a good way," Moses replied. "We must think about it. Change is difficult, but we must do something because we do not have enough food. I can see you have done many things to make your fields more productive. These are things we must do in Malenga. But I have another question. I see that you do not plow the soil. I remember you telling me before about that. Now I would like to know more about why you do not plow the soil."

Pastor Simon smiled. "Moses, you have observed something true; we don't plow our fields any more. This is a very important subject. It is time to have some tea. While we are drinking tea, I will explain to you why we no longer plow our fields."

Questions for study

1. *Why was it important for a field to rest between crops?*

2. *Why does Pastor Simon not allow his fields to rest for five years?*

3. *Why did God say the land must have a Sabbath rest?*

4. *Why are his fields able to grow good crops every year?*

5. *What do beans, groundnuts, and soya do to make the soil better?*

6. *How does alley cropping help crops grow better?*

7. *Explain how trees can help fields produce more food.*

Chapter 9

Mulching and Not Plowing

No more plowing

As they were drinking their tea, Pastor Simon talked to Moses. "Moses, when you were here before we talked about why we do not plow the soil. For many years we thought plowing was good. It made the soil loose so crops could grow easily. It allowed rainwater to go into the soil.

"Then we realized what plowing really does. It turns over the soil and exposes the deep soil to the hot sun. The hot sun quickly burns up any compost in the soil. It also destroys the good worms, insects, and other tiny creatures that keep the soil loose. The soil quickly became hard. It cannot absorb rain water well. All of this diminishes the fertility of the soil. So we thought that maybe it is not good to plow the soil. But we were not sure.

"A new thought then came to me. I remembered that God was the first farmer. He maintained the fields and forests long before he made man. God didn't plow the soil. The ground was always covered with plants and trees. God never turned the soil over. Maybe we should stop plowing and do as God did. So we stopped plowing and began what we call 'farming God's way.'"

Moses said, "That is interesting. Tell me how you do that."

Pastor Simon began explaining it to him. "God showed us we must follow certain principles carefully. The first one is:

1. Keep the soil covered at all times

"The hot sun burns the soil quickly. Heavy rains take much out of the soil if it has no cover. They take out nitrogen and other minerals the crops need. So we must keep the soil covered at all times. This includes the dry season and the growing season. That's what God did. He kept the earth covered with what I call 'God's blanket.' It is made of grass, bushes, and other plants. It is also made of leaves and branches that fall to the ground. We decided we should do the same.

"When we harvest our crops of millet, maize, and beans, we do not pull up the roots of the plants. We cut them off near the ground but leave the roots in the soil. We let the leaves or stalks of the plants lie on the ground to cover it. They become mulch, and this protects the soil. We do not burn the fields because that would destroy this fine mulch."

"Do the roots get in the way for planting the next season?" Moses asked.

"On the contrary," the pastor replied, "those roots are dead and make good compost. They are the place to plant new seeds because the soil there is loose. The dead roots help rainwater filter down into the soil. They help hold the soil together. This prevents erosion. So leaving the old roots helps the soil a lot."

2. Put a lot of mulch on the soil

"I told you before that we keep the soil covered with mulch. Mulch does five things:

1. It covers the soil and protects it from the hot sun and the heavy rains.
2. It decomposes and adds nitrogen and minerals to the

soil. You see, plants take things out of the soil to help them grow. Mulch puts those things back into the soil.

3. It keeps the soil cool, and little plants grow better.
4. It holds the rain water in the soil to keep it moist.
5. It keeps many weeds from growing.

John spoke up. “Did Jesus tell a story about mulch?”

“Yes, he did,” the pastor replied. “He told about a farmer who planted a new fruit tree. But after three years the tree did not produce fruit. He told his gardener to cut it down. The gardener did not agree. He said he would put mulch around the tree for a year to help it produce fruit. If the tree still did not produce fruit, only then would he cut it down. You see, Jesus understood the importance of mulch.” (See Luke 13: 6-9)

“How do you make all this mulch?” Moses inquired.

“We leave the stalks of millet or maize, and the plants of groundnuts and beans on the ground as mulch. When we cut or pull weeds, we leave the dead weeds on the ground as mulch. We add branches of the leucaena trees to make more mulch. We cut grass from other fields to put on top of the other things. We try to put down a thick layer of mulch, and the more mulch the better.”

“How do you plant through all that mulch?” Moses asked.

3. Planting where old roots are still in the soil

“As you can see,” Pastor Simon replied, “I plant my millet, maize, beans, and groundnuts in straight rows. That makes it easy to find the roots of last season’s crops. I pull aside the mulch to find the old roots. I make a small hole with my hoe right where the old roots are. I put the seeds in those small holes and cover them up. Then I bring the mulch back around where I have just planted the seed. In this way the soil is always covered. It is never exposed to the sun."

"Does that mulch keep the seeds from sprouting?" Moses asked.

"The mulch actually helps the seeds sprout. As I said before, it keeps the soil a bit cool by keeping the hot sun from burning the new little plants. The mulch helps the rain filter down better to where the seeds are starting to sprout. The water does not evaporate right away because the mulch is covering the soil. The new young plants can easily grow up through the mulch. And the mulch keeps weeds from growing rapidly around the new plants."

4. *Regular weeding*

"How do you get rid of weeds?" Moses asked.

"This is another principle God has shown us," the pastor replied. "We must keep weeds under control."

"Do you buy stuff to put on weeds that kills them?"

"No, I don't," Simon replied. "Weed killer costs money. It puts poison into the soil that can be harmful. That poison can also kill the crops we need. It is much better to pull out the weeds."

"How do you do that?" Moses asked.

"It takes a lot of work, but farming is work. Farming is good work because it gives us food to eat. Every two weeks I go through each field and either pull out the weeds or cut them off with my hoe. I put the dead weeds on the mulch to make more of it. I do this regularly, every two weeks, until the crops are thick and cover the soil keeping out the sunlight. That prevents new weeds from growing. Also, by weeding frequently, the weeds will not be able to produce seeds. Soon there will be fewer weeds.

"You see, Moses, we need to be faithful in our work just as God is faithful. By weeding faithfully, I have discovered something interesting. Each year I have fewer weeds than the year before. That is because only a few weeds go to seed. So only a few new weeds develop. Faithful work produces good results."

Simon weeding with a hoe. Virginia pulling weeds by hand.

"Do you do this weeding right up to harvest time?"

"Yes, and even beyond harvest time," the pastor replied. "In fact, weeding after the harvest, at the end of the rains, is very important. That keeps weeds from spreading and taking over the field between crops. Weeds need food and minerals to grow. If you let them grow, they will take food, water, and minerals your crops need. So it is very important to keep your weeds under control during the whole year. This will help save all your nutrients and water so you can have the best crop."

"Amazing," Moses said. "God has helped you learn many things. I had never heard such ideas, but they do make sense. Maybe we should stop plowing and do like you do."

"It would help you a lot," the pastor said. "And let me tell you more ways it will help."

5. *Less labor to prepare fields*

"Before I started using this kind of farming with mulch, I had to do much hard work. Every planting season I had to dig up my whole field. I had allowed all the weeds to grow up in the field since the last harvest. Some of those weeds had deep roots. They were very hard to get out. Often I would have much pain in my back. Now I dig only a little bit and it is just to dig holes to plant the seeds. My back is much better. It is easy to dig up small weeds with a hoe or machete when they are young. I save much hard work with this kind of farming."

6. *Plant crops as soon as the rains begin*

"Using the old way of farming, I remember we had to wait for the first few rains to come. Digging up the hard soil was difficult in the dry season. The rain helped make the soil soft so we could begin digging. We sometimes spent a week or more preparing our fields after the rain started. Now, we use this new way of farming with mulch and no plowing or digging. The fields are ready to plant without plowing. I can dig the small holes for planting seeds in advance. As soon as the first rains come, I plant my seeds. My harvest is better because my crops can use all the rain that falls from the beginning. I find this helpful because our rainy seasons are often more short than they were in years past."

John asked, "Uncle Simon, you keep cutting the branches of the leucaena trees growing around your fields. Why do you do that?"

7. Using tree branches for mulch

Pastor Simon said, "The leaves of the leucaena tree make good compost. I cut branches every two or three months. I lay the branches on the ground between the growing stalks of millet and maize, the groundnut plants, and the cassava."

Simon laying leucaena branches between the crops to make mulch.

"How does that help?" asked Moses.

"The fresh branches protect the soil from the hot sun and the heavy rain. The rain soaks better into the soil. The leaves cover the soil and keep it from drying out quickly. When the leaves die, they decompose and go into the soil to become compost. This makes food for the crops to grow better. So the branches help make the soil fertile. When the leaves are all gone, I cut more branches to put on top of the first branches."

Moses said, "In your border trees, I see small acacia trees growing between the big ones. Did you plant them?"

Pastor Simon replied, ''No, they have grown from the seeds that fell from the big trees. Acacia trees live only ten to fifteen years, and then they die. It is important to have tall trees around the fields to stop the wind. The little trees will grow up in the place of the trees that die. That way, we will always have a good wind-break."

Moses sat back in his chair and thought for a few moments. Finally he said, "You have given me many good ideas to think about. I must see how we can use them in Malenga to grow more food in the years to come. But my real problem is that we do not have enough food today. Planting trees to improve the soil will help us in a few years because trees take time to grow. But my people are hungry now, and our soil is worn out. What can we do now? Should we try to buy fertilizer?"

Restoring infertile soil

Pastor Simon replied, "No, I don't think so. Fertilizer costs much money. When you put the fertilizer on the soil, after a few rains, it is all gone. The rain carries it down deep into the soil. The roots of the plants can no longer get it. So it does not do much good."

"Then what can we do?"

Pastor Simon said, “Years ago our fields were like that. We had no trees. The soil was dry and sandy. It was infertile and we could grow little food. Then we learned about making Zai holes, and how they help us grow more food. Zai holes helped us a lot while we were making these other changes. Do you know what Zai holes are?”

“No,” replied Moses. “Tell me about them.”

Pastor Simon asked John, “Do they teach you about Zai holes in your agricultural school?”

“No,” John replied. “I never heard of them.”

Pastor Simon said, “Moses, meet me here tomorrow morning. We will go to the village of Kimweni where they are now using Zai holes. You will see how they are helping the people grow more food.”

Questions for study

1. *What does deep plowing do to the soil?*

2. *In what ways does a cover of mulch help the soil?*

3. *How can you make good mulch?*

4. *In the ‘no plowing’ method of farming, what are important principles to follow?*

5. *What are the advantages of not plowing fields before planting?*

6. *What did Pastor Simon mean when he said this kind of agriculture could be called ‘farming God’s way’?*

Chapter 10

Making the Soil Fertile

The next morning, Pastor Simon, John, and Moses, walked to Kimweni. Robert, the school director, and a couple of teachers from the school went with them. As they walked, Pastor Simon asked John to remind Moses that plants need food just as we need food.

"The best food to help crops grow comes from decaying plants," John said. "Dead leaves, grass, weeds, the stalks of maize, millet, and sorghum, and the plants or vines of groundnuts and beans can provide food for new living plants to grow. We call these decaying plants 'compost.' When there is little or no compost in the soil, crops do not grow well. The soil is infertile. When there is much compost in the soil, the soil is fertile and crops can grow well. Decaying plants spread on top of the soil is what we call 'mulch.' You saw mulch in Pastor Simon's fields."

Pastor Simon then explained the purpose of Zai holes. The purpose is to give compost to the growing crops. That way the crops grow better and produce more food.

When they arrived in Kimweni, they went to the house of Thomas, the head man. Pastor Simon introduced Moses, John, and the others to Thomas. He explained to him that he wanted to show Zai holes in the fields to Moses and John and tell them how they help people grow more food.

Thomas, the head man, was very pleased. He said, "We were very discouraged because our soil had become infertile. It was sandy and dry, and we had very little food. Pastor Simon came and showed us how Zai holes could help us grow more food. He showed us how to make them. Last year some of us made Zai holes and planted millet and maize in them. The stalks were strong and produced a lot of grain. This year, almost all of our people have made Zai holes. Come, let's go to my fields and I will show them to you."

Making Zai holes

As they walked to the fields, Pastor Simon told Moses how to make Zai holes. "At the end of the harvest, the farmer begins to dig Zai holes for the next planting. He digs a hole in the ground and put the soil beside it. The hole goes into the ground about 16 inches (40 cms), or the same distance as that between the elbow and the tips of the fingers. The diameter of the hole, or the distance across the hole, is also the same as the distance between the elbow and finger tips.

"The farmer makes lines of holes three feet (1 meter) apart across and up and down the field. In a field of about 2.5 acres (one hectare), which is 300 feet (100 meters) long and 300 feet wide, there will be one hundred lines of one hundred holes. That makes ten thousand Zai holes (10,000) altogether."

"Whew, that's a lot of work!" Moses exclaimed.

"Yes, it is a lot of work and it takes weeks to finish," the pastor replied. "Listen to what God says about that:

Lazy people want much but get little, but those who work hard will prosper and be satisfied.
Proverbs 13:4 (The Living Translation)

"During the dry season the people have little other work to do. So they have time to dig these holes in the dry season. They are then ready for the next planting time. That's what we did in Katindi for several years. We worked hard, didn't we, Robert?"

"We sure did," Robert replied.

"At the end of the harvest, the soil is still quite soft and the holes are easy to dig. After some weeks of dry season, the soil becomes hard. The holes are then difficult to dig. So we started digging the holes right after the harvest."

"What comes next?" John asked.

"When the hole is complete, we put into it grass, the leaves or stalks of millet, maize, groundnuts, or beans, or weeds that are dead. We would put 3 or 4 handfuls of dead plants or grass into each hole. We even put garbage from the kitchen into them, such as skins of bananas, papaya, or other fruits. We then put the soil back into the hole to cover up what we had put into the hole.

"This grass, dead plants, and garbage will decompose slowly during the dry season. It will become compost ready for the new crops. When planting time comes, the farmer can plant millet, sorghum, maize, beans, groundnuts, cassava, or squash in the hole."

Thomas added, "In fields where there are no trees, like what we had here in Kimweni, the wind will blow soil or dust into the Zai holes. This soil or dust may cover the holes completely. It will be difficult or impossible to find the holes when planting time comes. So I always leave a small stick, or a short piece of stalk of millet or maize sticking up in the middle of the Zai hole. That way, even if the wind filled the hole with soil or dust, I could still find each hole."

Digging Zai holes and filling them with compost and dirt. Marking their locations with small sticks to plant later.

The benefits of Zai holes

Harvest time was approaching. The millet and maize stalks were quite tall. The grain was almost ready for harvesting. Thomas showed them the ears of millet and of maize in his fields. He had planted all of his crops in Zai holes.

He said, "Look at these ears of maize. You see how big they are." Each one was big and had many kernels of maize on it. They looked at the millet. The ears were very big. Thomas put his hands around an ear of millet. He could barely get his hand around it for there was much grain on it.

He explained, "The compost in the Zai holes helps the crops grow better. Also, the Zai holes hold on to rainwater and prevent the heavy rains from washing away a lot of soil."

They walked to another man's field. This man had not made Zai holes in his field. They looked at the maize and millet in that field. The ears of maize were small. They did not have many kernels on them. The stalks of millet were short. The ears of millet were about the size of Thomas's thumb. They were small, not like those grown in Zai holes.

"You see," said Thomas, "Zai holes help us very much. Last year I had enough millet to last all year. Our family will have even more millet, maize, and groundnuts this year. The holes help us very much."

"But it takes much work and time to make all those holes, doesn't it?" asked Moses.

"Yes, of course," replied Thomas. "If we want to have much food to eat, we must work hard. The Bible says it is by the sweat of our brow that we get our food, doesn't it, Pastor?"

"Indeed it does," replied Pastor Simon. "We must work hard, and work in the way God wants us to work. When we follow the right methods of using the soil and caring for it, we can have much food. We will not be hungry."

Moses asked Pastor Simon, "You do not use Zai holes any more. Why not?"

Zai holes no longer needed in Katindi

Pastor Simon answered, "In Katindi, we used Zai holes for four years while the trees were growing up around our fields. After three or four years, the trees were putting nitrogen into the soil. The branches from the leucaena trees began making good mulch all during the growing season. We found that it was not necessary to use Zai holes any more. Our fields now give much food."

John spoke up. "I think I see what's happening. In the old days, your fields and gardens did not produce much food. You saw the soil was not fertile. You learned about compost and decided Zai holes were a good way to get compost to your crops. So the Zai holes were a temporary means to improve the soil. Then you realized border trees and mulch would be much better. As the border trees grew and you used mulch to keep the soil covered at all times, you no longer needed Zai holes. Is that right?"

"Exactly," Pastor Simon replied. "Zai holes are good to use between the old way of farming and the new methods of farming we now use in Katindi. They are good to use while waiting for the leucaena, acacia, and grevillea trees to grow around the borders and in the fields. That is what Thomas is doing here."

Moses asked, "Do you think we could use Zai holes in our fields?"

"Of course," replied Pastor Simon. "You must explain to the people how to dig them, starting right after the harvest. Show them how to put dead grass, plants, leaves, and garbage into each one. This will take much work. But it will help them grow more food. You can bring your people here to Kimweni to see the people here digging them. Thomas will be glad to show your people how to do them, won't you, Thomas?"

"Certainly," Thomas replied. "We are always happy to show people from other villages good things we have learned. That is what you taught us to do, Pastor Simon."

Pastor Simon added, "You must also tell your people, Moses, why it is important to let some trees grow in their fields. You can show them how to plant acacia and leucaena trees around the borders of your fields. If they start doing this during this season while they are digging and using Zai holes, after four or five years your fields will again be fertile. You also will no longer need Zai holes.

"Look at the young acacia and leucaena trees growing around the fields of Thomas. Soon they will be big and he will no longer need to make Zai holes."

John said, "Uncle Simon, one big advantage of all of these methods using trees and Zai holes is this. It does not cost money. It takes much work, but people have time during the dry season to do that work. So they do not need to spend money for expensive fertilizer. That is good because many people do not have money to buy it."

"That is correct, John. This is why we are teaching people in many communities how to use these methods. People from these communities are now teaching it to people in communities further away. Slowly more people are learning these methods. This will help us to restore our land."

Compost

The men walked back to the house of Thomas. They sat in the shade under a large spreading bougainvillea bush. Thomas' wife Sarah brought them tea. As they sat drinking, other men and a few women from Kimweni came and sat down. Sarah spoke up.

"We try to have a garden of vegetables near the house. But the soil is so poor almost nothing grows. What can we do?"

Pastor Simon turned to John and asked him to explain to the women of Kimweni how to make compost for their vegetable gardens.

John began describing to them how to have a good supply of compost always available for their gardens. "You make compost from dead plants, like weeds and the old plants of your food crops. Dead leaves also make good compost. These dead plants and leaves rot but have many minerals in them that growing plants like vegetables need.

"First, find a spot in the shade near your house about 12 feet (4 meters) by 6 feet (2 meters). Build a small shelter over it to keep out the hot sun and heavy rain. Put a fence of sticks and slats around 3 sides of it. Divide it into 3 sections by putting a small fence between each section.

"Begin to collect dead plants and leaves and put them into the first section. Try to find some animal manure. If you have chickens, rabbits, or guinea pigs, put their manure onto the pile. You can use goat manure or even a little cow manure as well. The manure helps the dead plants and garbage rot faster. You don't need much. A thin layer of manure on top of the pile when it is about as high as the length of your forearm is enough. If you use a lot, it can burn the dead plants.

"You can put garbage from the kitchen onto the pile as well, like the skins of bananas, papayas, or citrus fruits. You can also use stalks of maize and millet, and even the ears after you have taken off the grain or kernels.

"The compost needs water from time to time. If it does not rain, put a bucket of water on it at least once a week. The water helps the plants rot well. When the pile gets about two feet high (1/2 a meter), take a shovel and turn it over. What was on the bottom and is starting to rot is now on the top. Keep adding more dead plants to it.

"When the compost in the first section is 4 or 5 feet high (1 1/2 meters), take a shovel and move it into the second section. It will continue to decompose there. Start a new pile in the first section. When that first section is full, move the compost in the second section into the last section. Put the compost in the first section into the second one. And start a new pile in the first section. By that time, the pile of compost in the last section, that started in the first section, is ready to use."

Three sections makes a good compost pit.

"How do we use it?" Sarah, Thomas's wife asked.

"In your vegetable garden, make seed beds. They should be raised about 8 inches (20 cm) above the ground. You can do this by putting logs, or short lengths of bamboo, around the seed bed. Take the soil out of the middle of the seed bed. Put a lot of the compost from the third section in the middle of the seedbed. Then put the soil back on top of the compost. You are now ready to plant your vegetables."

Another woman asked, "Are you sure the vegetables will grow?"

Putting compost into seedbeds.

"Sure," John replied. "That is, if you water them regularly. You must have a good fence around the vegetable garden to protect it from chickens, goats, and other animals."

Pastor Simon spoke up. "You can also use the compost in Zai holes if you have enough of it. Or in the holes where you want to plant new trees. The compost helps the crops or new trees grow well."

He opened his Bible and read to them the parable Jesus told about the man who had a fruit tree that was not producing fruit.

> *There was once a man who had a fig tree growing in his vineyard. He went looking for figs on it but found none. So he said to his gardener, "Look, for three years I have been coming here looking for figs on this fig tree, and I haven't found any. Cut it down! Why should it go on using up the soil?" But the gardener answered, "Leave it alone, sir, just*

one more year. I will dig around it and put in some fertilizer. Then if the tree bears figs next year, so much the better; if not, you can have it cut down. Luke13: 6-9 Good News Bible

Moses spoke up. "Last night I thought about that story you told us yesterday. I was impressed that even Jesus understood the importance of providing food in the soil for trees and crops to grow. So, if we want to grow good vegetables and have good crops, we need to make compost for our gardens and put a lot of mulch on our fields."

Fertilizer

Thomas spoke up. "Pastor, the parable you read talks about fertilizer. Should we put fertilizer on our gardens and fields?"

His wife immediately asked, "Excuse me. What is fertilizer?"

John replied, "Plants need certain things in order to grow and produce food. They need water, air, and sunshine. The green substance in the plants, called 'chlorophyll,' uses water and air to make the plant grow. The sun gives the energy for this. However, the plants also need certain minerals in order to make chlorophyll so the plants can grow. This requires nitrogen, phosphates, and other minerals.

"The fertilizer you buy contains these minerals. The fertilizer factory gets these minerals, mixes them together in powder form or in small lumps, and puts them into sacks to sell. We call this 'chemical fertilizer,' for the minerals come from chemicals. You can buy a sack of chemical fertilizer and put it on your fields at planting time. When the new plants grow, their roots will take these minerals from the fertilizer in the soil. This helps them grow fast and big and produce more food for you."

Pastor Simon said, "John, you have given a fine explanation of what fertilizer is and what it does. Sarah, do you understand this?"

Sarah answered, "Yes, I understand it. It sounds good because I need much food to feed my growing children. But we are poor, and we do not have money to buy fertilizer."

Pastor Simon said, "Sarah has mentioned a real problem about fertilizer. It does cost money, and many poor people cannot buy it. In many places, you cannot find fertilizer in the stores to buy. Using fertilizer has other problems as well."

John said, "Uncle Simon, can you tell us what they are?"

The pastor said, "People who use fertilizer must have a good knowledge about it. They make different kinds of fertilizer. To know which kind to use, you first have to test the soil in the fields and garden. It costs money to buy the means to test the soil. But without knowing about your soil, you won't know which fertilizer to buy.

"You must also know how much fertilizer to use. You need to know when to put it in the soil, and the best way to do so. If you put much fertilizer by the plants, it will burn them and they will die. But if you put only a little in the soil, it will not help the plants. So you will waste the money you spent on it."

Robert spoke up. "Pastor Simon, there is another problem. My brother lives in another country where they grow much food. Most of the farmers use a lot of chemical fertilizers. But now the people there cannot drink the water from their springs or wells. The chemicals go down into the water table and make the water impure. It is no longer safe to drink that water for it can make people sick. They now have to buy drinking water in bottles."

John said, "We must remember something else about fertilizer as well. It works quite well on soil that is heavy, that has a lot of clay or plant matter in it. The minerals stay in that soil for some time. The roots of the new plants can find it and use it to grow well to produce food.

"However, our soil has much sand in it. If we put chemical fertilizer in sandy soil, the sandy soil does not hold the minerals in the fertilizer. The heavy rains quickly take the minerals deep down where the roots of the new plants cannot find it. It will be gone before the new plants can use it. So in our sandy soil, we may lose much of the fertilizer we put on the soil."

Pastor Simon concluded, "Sarah, you see that making compost and using it in your garden and even your fields is better than buying fertilizer."

Robert spoke again. "Pastor Simon, I am a teacher. When I come to the end of a lesson, I give a short summary of what I have taught. I have listened to this talk concerning fertilizer and compost. Can I make a summary of what I have understood? I fit is not right, you or John can correct me."

"That's a fine idea, Robert. Go ahead."

Robert said, "We have two ways we can make the soil strong. One is to buy chemical fertilizer and to use it correctly on our fields. This will cost us some money. But we will get more money from the extra food it helps us to grow. It may also make our drinking water bad.

"The second way is to make compost and put it into the soil. Compost comes from the plants all around us. It costs us nothing, just extra work. The minerals in the dead plants provide minerals the new crops need to grow well. Have I spoken well?"

Pastor Simon and the others all agreed with what he said. Pastor Simon added that God has given us many good things in the soil and plants. He has also given us our intelligence. If we learn how to make good use of what God has given us, and work hard to use these resources well, we will be blessed by having good food.

Before the people left to return to their homes, Robert prayed. He thanked God for what they had learned. He asked God to help them to make good use of this new knowledge. He asked for wisdom in how to use correctly the many good things God makes available to us in nature, in the soil and the plants.

The problem of erosion

As they walked back to Kalindi, the men talked more about trees and fields. They walked up a hill and through the town of Namuru. As they looked at the small farms around Namuru, they could see many gullies made by the rain. Pastor Simon pointed these out to Moses. He said this was a big problem in many communities. He had already sent word to farmers and church leaders in many communities in the area, including Namuru, to come to Kalindi the next Saturday to talk about the problem of erosion and what each farmer needed to do. Moses said that he would come with some of his farmers who also had farms on the hillsides. Erosion was a big problem for them.

When they came to the crossroads, they said goodbye, and Moses thanked Pastor Simon, John, and the others for all they had told him. He said he and some of his farmers would come on Saturday, for they really needed to learn about erosion. After a brief prayer, Moses turned and went down the road to Malenge, and the pastor, John, and the others returned to Kalindi.

Questions for study

1. *How big and how deep are the Zai holes?*
2. *How far apart should you make them?*
3. *What is the purpose of putting dead plants, weeds, and even garbage from the kitchen into the Zai holes?*
4. *Why do crops planted in Zai holes grow better?*
5. *Is this something you can do where you are?*
6. *What can you do so that, in a few years, Zai holes will no longer be necessary?*
7. *What steps must you follow to make good compost for gardens?*
8. *What are the things you can use to make compost?*
9. *Which is better: chemical fertilizer or compost?*
10. *Read Genesis chapters 1, 2, and 3. What do these chapters say about work?*

Chapter 11

Stopping Erosion of the Soil

A large group of farmers and church leaders living on the nearby hills and mountains gathered at the church in Katindi. They had come to talk with Pastor Simon and other church leaders about the problem of erosion of the soil on the sides of the hills. During the heavy rains, they lost much soil from their fields and hillsides. The soil lost was the top soil. This is the best for growing crops. Their fields were growing less food because the deeper soil that remained was not fertile. Many gullies and deep ravines had developed and they were getting bigger. Some farms had even fallen into the ravines and been lost. The problem was serious.

They all recognized that Pastor Simon had become wise in how to grow much food and keep the land strong and healthy. His nephew John had come from the agricultural school to help with this meeting. The people were eager to learn how they could protect their little farms and grow more food.

God tells us to rule over the land

Pastor Simon began the meeting by reading from Genesis chapter one. He reminded the people that God has put us in charge of the land, the forest, the soil, and the animals. We are to rule over them; they are not to rule over us. "However," he said, "to rule over them well we must do two things. We need to seek and follow the wisdom of God. We also need knowledge about how to rule over our fields and forest. That is why we have come to this meeting today."

He prayed and asked God to help them in their discussion. Then he asked them to tell about what was happening in their fields and gardens. He wanted to hear from them what their problems were. Several of them described gullies that the rainwater had formed through their fields.

Planting trees around fields

"Many of your fields are on the sides of the hills," Pastor Simon said. "Do you have trees planted around each field?"

Most of them said they did not have trees around their fields. The pastor then asked John to explain why planting trees around every field is important. John explained how a row of trees planted closely together helps hold the soil and stop erosion. He told how certain trees, like leucaena and acacia, put nitrogen into the soil to make it fertile.

Up-and-down fields cause erosion

The pastor asked them if their fields go up and down the hillside, or if they go across the side of the hills. Many of the farmers said their fields go up and down the slopes.

"This is a real problem we must talk about," Pastor Simon said. "Fields should go across the sides of the slopes and not up and down them. If you plant your crops in rows going up and down the hillside, the rainwater will flow down between the rows and take away much soil. Soon gullies will form and will become bigger with each rain."

Several of the farmers nodded their heads. "That is exactly what is happening on our farms," they said. "How can we know how to make rows across the slopes instead of up and down?"

A field of maize planted up and down a hillside. Gullies made by rainwater between the rows.

Making rows go across the slopes

John now spoke up. "You need to make a simple instrument to show you the level of the ground. We call it an 'A frame' because it looks like the letter 'A'. Take two poles, or lengths of bamboo, equal in length and about 5 feet (1.5 metres) long. Tie or fasten them together at one end. Find a level place and spread the other ends apart about one meter. Now tie or nail another shorter pole or piece of bamboo across the two poles.

Put it half way between the ground and the top where the poles are fastened together. It will now look like the letter 'A'.

"Next, tie a string at the top of the frame so that it hangs down in the middle. Tie a small stone or other weight on the lower end of this string, about 3 feet (a meter) below the top. The string should hang down across the horizontal bar.

"Place the A frame on ground you know is level. If you have a carpenter's level, use it to be sure the ground is level. Where the string crosses the horizontal bar, make a mark. Turn the A frame around and make a mark on the other side of the bar. The two marks should come in the same place. If they don't, make a final big mark halfway between the two marks."

John had an A frame there and showed it to the group. They could see that it was very easy to make.

Note the string hanging down across the middle of the A frame cross bar.

John set the A frame on the level ground in front of the group. The string with the weight hung down from the top and crossed the horizontal bar right at the big mark. “This means that the ground here is level,” he said.

He next set one pole of the A frame on a small step; the other end was on the ground. The string now hung off to one side, several inches (or centimeters) to one side of the big mark. “This shows that the A frame is not level,” John said.

“You can make an A frame like this for yourselves,” he said. “Take it to one of your fields. Set it up on a slope on the top side of the field. If the string hangs down on one side of the mark on the horizontal bar, you know the A frame is not level. The ground in that direction is going up and down. Now move one end of the A frame around until you find the place where the string hangs down across the big mark. You know that the ground in that direction is level.”

Again he demonstrated how easy it is to do this. “Move the A frame in the level direction along the side of the hill. Put in small stakes as you go to make a level line. When you get to the end of the field, you know you have gone along the side of the hill and not up and down.

“Now move down the hillside 3 or more feet (one or more meters) and go back along the side of the hill to make another lower level. Put in little stakes as you go. Keep going back and forth along the field in this way to make more level lines. Make them across the field starting at the top. Go all the way down until you come to the bottom of the field. You now know that you have level rows along the sides of the slope and not up and down. Make your rows along these level lines. That will keep the rainwater from washing away the soil.

Pastor Simon spoke up. "This is what we call 'contour farming,' farming along the 'contours' of the slopes. It is essential for stopping erosion."

Measuring contour lines on a hill with A frame. Strips of grass being planted on contour lines like those on the hill in the distance.

Making terraces across fields

He turned to John. "John, some of these farmers have quite big fields that go down the slopes. Explain to them about making terraces."

John explained that terraces provide contours on the field that can last for many years. "Make a level contour line across the field 10 to 15 feet (three to five meters) from the top. Make more level lines every 10 to 15 feet (three to five meters) below the previous lines. You now have a series of level or contour lines across your field three to five meters apart.

"Plant a narrow strip of thick grass like vetiver grass, about 50 inches wide (20 cm), along each line. This will make a thick wall of grass across the field on each line. Each time it rains, the water will carry a bit of soil from the upper side of each terrace down to the lower side. The snip of vetiver grass will hold it there. After a few rains the terrace will become quite flat. You can now plant your crops along each terrace. There will be no erosion."

Pastor Simon explained how flat terraces permit the rain water to go down into the soil. It will no longer run off down the hillside and cause erosion.

John added, "Sometimes rainwater collects in the terraces. This is true where the soil is hard or thick and the rain goes into it very slowly. If a lot of water collects, it may break through the grass strips and make a big gully. So where it collects, make a small dam of earth to hold the water there. Some of it will go into the soil. The rest will evaporate."

Making the soil better

The pastor described how mulch can protect the soil from erosion. They should lay down branches from the border trees of leucaena or acacia between the growing plants. The leaves of those branches will dry up and decompose. They will become mulch to make the soil rich. They will also protect the soil from the hard rain. Hard rain on unprotected soil leads to erosion. The mulch lets the rain water go down into the soil instead of running off and making erosion.

"I want to say one more thing about planting crops," the pastor said. "We have learned that plowing the soil deeply is not good. It exposes the soil to the sun. The sun quickly burns up the compost in the soil. Deep plowing also lets heavy rains take away much soil by erosion."

"How do we get our fields ready to plant?" one of the farmers asked.

"You should simply break up the soil with a hoe," Pastor Simon replied. "Leave the dead plants and roots from the previous harvest in the soil. They will be food for the new crops to help them grow well. Break the soil up gently but do not turn it over. To plant, make little holes in which to put the seeds. The young plants will grow better. Keep the soil covered at all times with mulch. That way the rain will not wash away much soil by erosion."

Getting rid of gullies

One farmer asked, "Pastor, I already have gullies through my fields. What can I do to fill them in and keep the rainwater from making them big?"

"The first thing to do is to start at the top of each gully," Pastor Simon said. "Try to stop rainwater from going into the gully at the top. Make a small dam of earth at the top to prevent water from going into it. Plant crops or vetiver grass along the sides of the gully. Start at the top and go down both sides of it. You may even be able to plant a cover crop like stylosanthes down in the gully, along its sloping sides. This will hold the soil in the gully and stop the rain from washing it away."

John now spoke up. "You also need to do something at the bottom of each gully. Put three or four lines of small stakes or poles across the gully. Then put a lot of brush or grass on the ground just above them. This will help hold the water and also the soil it washes down the gully. If you have stones in the field, make a pile of them across the bottom of the gully. The stacks of brush or stones will help the gully fill up with soil."

The pastor said. "If the gully is long, you will need to make lines of stakes or stones in several places. Each one will help hold the soil the rain is washing down it. They will slow down the flow of the water. Now it won't wash away much soil."

John added, "If the sides of the gully are steep, this leads to much erosion. Use a shovel to push the dirt on the top of each side down into the gully. You want to make the slope of the gully sides shallow. That will keep water from flowing fast down the gully. Then plant grass or even crops along the sides of the gully. These will help hold the soil. It will keep the gully from becoming very deep."

Metal roofs on houses

Pastor Simon now mentioned another problem that can cause soil erosion. He asked the group, "How many of you live in houses with metal roofs?"

Most of the men said they did.

"When it rains, what happens to the water that runs off your roofs?" he asked.

"It runs across the yard and then down the hill below our town," several of them said. The pastor commented that this is a lot of rainwater causing more erosion.

"What can we do about that?" they asked.

Grass around the edge of the yard and foundation of the house holds water and prevents erosion.

"Make sure the ground in your yard is level. Plant strips of thick grass, like vetiver grass, all around the sides of your yard. Vetiver grass has deep roots that hold the soil. Also, plant vetiver grass around your house right next to the foundation. The grass will keep water from eroding soil around your house. It also will help keep termites away. Keep the grass cut short so snakes won't live in it. If you have small ravines running down from your house, fill them and plant grass there. It is important to stop all rainwater from rushing down the hill. It takes much soil with it."

The problem of big ravines

Another farmer spoke up. "Pastor, some of the ravines on our hills are deep. They have been there a long time. What can we do about them?"

Pastor Simon replied, "These ravines show that we have neglected this important issue for a long time. We have not done what God has told us to do, to care for the land and protect it." He pointed to the hills in the distance. The group could see several deep ravines coming down the sides of the hills.

"Those ravines," the pastor said, "have destroyed much good land on those hills. They allow lots of rainwater to run down to the valley instead of going down into the soil where it is needed. This often causes floods which damage farms in the valley. They bring down much soil from erosion into the rivers in the valley. This soil can kill many fish in the rivers."

A big ravine caused by erosion.

"What can we do about them?" the farmers asked.

"The first thing to do is to start at the top of each ravine. Do what I told you to do with the gullies," the pastor replied.

"Try to dig a channel on each side to divert the water. Or build dams of earth at the top so the rainwater will not start to go into the ravine. Plant grass immediately on the sides of these channels or dams to hold the soil. If the grass grows quickly and is not washed away, these channels or dams will help for a long time.

"The next thing is to plant fast-growing grass or trees along the sides of the ravine. This will help hold the soil in place. It will also keep rainwater from running down the sides of the ravine and then going down the middle of it."

"What can we plant?" they asked.

"Bamboo, other fast-growing trees, or else grass like vetiver," the pastor replied.

Restoring a gully by building dams to divert water runoff, planting grass and trees, and filling deep sections with barriers of rocks and sticks.

He continued. "These ravines are also a good place to begin planting slow-growing trees of the forest. It will take several years for the trees to grow tall. Once they are tall, they will hold the soil and eventually will fill the ravines with mature forest. This will bring birds and small animals back to our land."

"We don't know anything about planting trees like that," the farmers said. "What do you mean by planting trees of the forest?"

"That is a very important matter as well," the pastor replied. "Let's have another meeting next month and talk about this. I would like you all to come. We can restore the forest to the parts of the hills where we cannot plant crops. If we do this, both the hills and the valleys will be much better. Even the air we breathe will be better because the big trees of the forest help purify the air."

The members of the group thanked Pastor Simon, John, and the other church leaders for this meeting. They each said they had learned much, and planned to come for the meeting about planting forest trees. Pastor Simon reminded them that caring for the land is the responsibility God has given us. When we work hard to do that, we are doing God's work. He can help us with this if we ask for his wisdom and work together to do this. They concluded the meeting with prayer. Several of the men thanked God for his help in this discussion and asked him to guide them all in restoring the land, the fields for crops, and the forest.

After the group left, Robert stayed to talk to Pastor Simon, John and Moses. He said, "We have another serious problem we must discuss also. In most of our rural communities people have only one way to make money. It is by using the land either for farming or for raising animals. This is why our soil has become weak and infertile. Until we find other ways people can make money, we will not be able to restore our land. Pastor, I suggest we spend a little time thinking about ways people can make money that do not depend on the land."

Pastor Simon smiled and said, "Robert, you are absolutely right. This is indeed an important subject. But we have had a long day. We have seen many things and had a long talk. In the morning come to the house and we will drink tea together. Then we can talk about how people can earn money, improve our local economy, and make our communities prosperous."

Questions for study

1. *Why is it important to plant crops in contours along the sides of slopes instead of up and down them?*

2. *How can we make level contour lines along hillsides?*

3. *What do we need to do to make permanent terraces on hillsides?*

4. *How can we restore gullies already formed so they become part of fields for crops?*

5. *What can we do in deep ravines to keep them from getting bigger and continuing to cause floods?*

6. *Why should we do all of this work? What benefits will we get from it?*

Chapter 12

Ways to Earn Money in a Community

Why the forest is gone

The next morning, after drinking tea and eating groundnuts, bread, and fruit, Pastor Simon, Moses, and John sat and talked. The pastor asked Moses to tell the others why the people of his village had difficulty growing enough food. Moses thought for a little while, and then he began talking.

"Many years ago we had a big forest. Animals and birds lived in the forest, and we could often eat meat. Mushrooms grew in the forest and many kinds of fruit. Our ancestors passed this forest on from one generation to the next. They said no one should ever destroy that forest.

"But then the village began to grow. In the past many of our children died, but now, with new medicines, many of them lived. So we needed more food and more land to grow food.

"Our children wanted to go to school, and that costs money. They needed clothes, and that also costs money. We realized we had to have more money. But the only way we could get money was to grow more food and sell it. To do so, we needed bigger fields and gardens. So we began cutting down the forest to have more fields."

John spoke up. "You are saying that as the village grew, the forest became smaller because you used it for fields to grow food."

"That's right, "Moses replied. "In the old days, we planted a field and harvested the crop. Then that field would rest for five years before we would plant it again. In this way, the soil in the field became strong again. We don't do that now. We need more food for ourselves and more food to sell for money. So we continued to cut down more forest. We also could no longer let a field rest for five years. We now plant our fields every other year and sometimes every year."

"What happened to your fields?" the pastor asked.

"We say the land has become tired. It cannot give us much food any more. Each year the fields give us less and less food. But we do not know why."

Why the land is tired

John said, "I can tell you why. Plants need food in order to grow well. They find the food in the soil. The food comes from plants and grass that have died. It comes also from the leaves of bushes and trees. The plants and leaves fall onto the soil, decompose, and become part of the soil. This provides food for the new plants or crops. Other food in the form of minerals comes up slowly from deep down in the soil. It is the roots of bushes and trees that bring this food up from deep in the soil.

"When you plant the same field every year or every two years, the bushes and small trees cannot grow. They are not there to drop leaves onto the soil or to bring minerals up from deep in the soil. So the soil has no food in it.

"Because the soil has no food, the crops you plant cannot grow well. They cannot produce food. You are right when you say the soil is tired out. You could also say the soil is hungry, even starved. So it cannot produce much food."

Pastor Simon spoke up. "Moses, it is like asking a man who is hungry and tired to keep on working. He can't until he has had a rest and good food to eat."

John said to Moses, "That is what you are doing to your soil. You are making it work almost every year. It cannot get the food it needs to feed the crops you plant. The soil is starved and tired out. It can no longer produce the food you need to eat."

The land is the only way to make money

"But Pastor, "Moses said. "We have to grow a lot of food. We need to sell the food to get money. We have to send our children to school. We have many things we want to buy. The land is the only way we can make money."

The men were silent for a moment. Robert had joined them. Pastor Simon said. "Moses, this is the real problem Robert brought up last evening. You do indeed need money. We all need money. Like you say, using the land to grow food or to raise animals is the only way to make money. On the other hand, the land is getting tired, so it can produce less food to sell for money. It can support fewer animals to sell."

"So what can we do?" Moses asked.

"This is an important question," the pastor replied. "We need much wisdom as we think about it. The answer is quite simple, yet it means we must change how we think about getting money. We must find other ways to make money that do not use the soil."

"What do you mean?" Moses asked.

The pastor turned to Robert and asked him to tell Moses what happened in their school a few years ago. Robert smiled and began talking.

Making money in other ways

"One day I asked a class to list the things they often bought at the store in town. We looked at the list together to see if we were buying things that we ourselves could make and sell. Soap was on the list, so I told two of the boys to read how to make soap. They did so and discovered what is needed to make soap. I assigned them a project to buy some oil and caustic soda, find a couple of big pots, and try to make soap. The science teacher helped them, and they made several bars of soap. They then began making soap each weekend and during vacations. They have now finished school and are making soap every day and people are buying it. They have a small business here, with other people working for them. They are making good money."

He held up a bar of soap and said, "Here is a bar of soap they have made. You can take it home with you and use it."

Moses was astonished. "They make bars of soap like this right here in Katindi?"

"That's right," Robert said. "They even put a bit of perfume in it, and it smells nice."

Robert went on. "On that list were clothes for babies and little children. I talked to Virginia here, and she began showing a few girls how to make them. Three of them are now making and selling them from their homes. People come from other towns nearby to buy them."

Women make bread

Woman baking bread to sell. Her daughter is helping her.

Yvonne spoke up. "Some of our women decided to work together to make and sell bread. People are happy to buy fresh bread. At the beginning ten of the women put together some money and bought a sack of flour. They gave it to one woman and she started baking bread to sell. The second week, they gave more money and bought another sack for a second woman. They kept on doing this for nine weeks, so each woman had flour to make bread for nine weeks.

The tenth woman is in the market every day to sell the bread the others have baked. The money is then divided among them all. From what they earn, they have enough money to keep on buying flour and to support their families."

Woman selling bread and bars of soap in the market.

"You see, Moses," Robert said. "A lot of things we buy from stores we can make ourselves. I am sure there are people in your village who could begin making things to sell. You probably have some young men who could learn simple carpentry. They could make and sell tables, chairs, and other items of furniture people need."

Roger spoke up. "That is exactly what I did, and now I am making a good living. My wife and I have a small garden to raise vegetables. But we do not have fields to grow maize or millet or groundnuts. We buy them from our farmer neighbors."

Roger making a living by carpentry.

Moses seemed unsure of this, and he said. “Our young people don’t want to learn a trade. They want to finish secondary school and go to the university. Then whey go away and leave us.”

Pastor Simon spoke up. "Those that can finish secondary school and go to the university should do so. But you know that others don't do it and drop out of school. They are the ones who could learn a trade like Roger did and start making good money."

"How is that going to help us have more food?" Moses asked.

Community businesses will help grow more food

Robert answered, "It will increase the money in your village. Let's say that half of your people make money in ways other than growing food. They won't need fields to grow food. Their fields could become forest or places to raise animals. The other half can use their fields to grow enough food for themselves and to sell to the people doing these small businesses. You will no longer sell your food at low prices to merchants who then take it to the city and sell it for high prices. Your farmers will sell food at better prices to the people in your village who are making money in these other ways. All the food you grow, and the money you get from it, will stay in your village. Everyone will profit from this, and the land will get better."

Robert added, "The son of one of my friends has learned how to repair bicycles. People are always bringing him their bicycles to repair. He makes good money doing that.

Rebecca spoke up. "Arthur, a member of our church, went to a vocational school run by a group of churches in the district center. He is now a mason and helps people build their houses. He makes a lot of money."

Pastor Simon said, 'These are good examples of how people can make money by making things or repairing things. They learned how to do this, and they now make money in ways that do not hurt the soil."

Making money by repairing bicycles and shoes in a community.

Young people will stay in the community

Robert added, "If we can help our young people find ways to make good money right here in our villages, they will stay. They will not go away to the big cities to look for jobs. They will make money to support themselves and their families. They will also buy the food the farmers in the community grow. So this will help everyone in the community live better."

Pastor Simon said, "Let us think about it this way. Life in a community is like a table. A table has at least four legs. One 'leg' is growing food. If you have a carpentry shop in your community, that would be a second leg. Perhaps someone could learn how to make candles. That would add another leg. Someone else might learn how to make leather out of the skin of cows or goats, and that would be another leg. And so on. Every community needs to have people who make things to sell, or can repair things people have. Life in such a community would be like a strong table that has many legs. It can support a heavy weight because it has many legs. But in your community, Moses, you have only one leg on your 'table.'"

Moses thought about this for a moment, and then said, "I think I understand what you are saying. But how can we get a carpenter shop in our community, or someone to repair bicycles?"

Pastor Simon smiled. "Talk this over with your leaders, your school teachers, and with the women. Teachers can give students projects of how to make things to sell, like Robert did here. Women can bake bread or make clothes like they do here. Other young people can learn trades from older men or women, or even go to a trade school nearby to learn how to be a tailor, a mason, or a mechanic. They will soon be able to make or do things to make money. This will help your whole community."

David, the farmer, had joined the group. He now spoke up. "My grandfather had a nice little farm of ten acres. He could grow all the food his family needed. He had three sons. Each of them married, and my grandfather divided his farm between them. My father got three acres. He had trouble growing enough food for us. He had two sons and finally divided the farm between the two of us. I have only an acre and a half and cannot grow enough to feed our family of five and have enough to sell for the money we need. So I have learned how to make simple farming tools and sell them to others."

"What are your children going to do?" Moses asked.

"I already have a son in carpentry school," David replied. "His younger brother wants to learn computers and get a job. I want my daughters to learn how to be teachers or secretaries. So none of them will need to live off the land."

Robert asked him, "Don't any of them want to go on to study in the university?"

David answered, "We have talked about that and they decided not to. Many who finish the university cannot get good jobs. Some go back to the farm, and others just sit waiting for a job. My children believe it is better to go to a school that can teach you how to make your own money."

"That's a smart idea," Robert replied. "I wish more young people would do that."

Moses said, "This is a new idea to me, and I am sure it is new to my people. I don't know if they will accept it."

A strong community

The pastor replied, "Actually it is not a new idea. In fact, it is a very old idea. We find it back in the beginning of the Bible, in the book of Genesis, chapter four. This tells us how men, at the beginning of human life on earth, learned different trades. Some learned to raise cattle; others to grow food. Some were blacksmiths and made tools out of iron. Some learned how to make instruments of music. Some knew how to build buildings, and cities grew. So today we need to learn how to do different things. Each person will do something other people want. He will then get money from other people for the things he does. In this way, the whole community will grow and become prosperous."

Robert added, “Some will still plant crops and sell them. They will take care of the soil so that it remains fertile like we do here in Katindi. They can make money by selling food to teachers, carpenters, masons, repairmen, and all the others.”

Moses was silent for a few minutes. The others could see he was thinking about something. Finally he spoke up and said, “I am the chief of Malenga. The number of people in our town has grown a lot. We no longer have enough land to grow crops. Even I do not have enough land for my family. What can we do?”

“Moses,” the pastor replied. “That is a big question. We have discussed it here in Katindi. The question is how can we plan our families and our land so that everyone will have enough to eat and ways to make money. I will call some of our people together, and this afternoon we will talk about it.”

Questions for study

1. *What things lead people to cut down more and more of the forest to grow more food?*

2. *List several reasons why the soil becomes tired and unable to produce much food.*

3. *Why is it important for people to learn how to make money in ways other than planting crops or raising animals?*

4. *Make a list of the things you could make to sell to others in the community in order to make money.*

5. *Are there places your children can go to learn trades so they can earn their own money?*

6. *The Bible says: "Make it your aim to live a quiet life, to mind your own business, and to earn your own living, just as we told you before. In this way you will win the respect of those who are not believers, and you will not have to depend on anyone for what you need" I Thessalonians 4 : 11 and 12. What is God teaching us in this passage?*

Chapter 13

Planning Our Families

That afternoon the pastor and Moses sat with a group of people in a circle under the mango tree in Pastor Simon's front yard. Robert, David, Yvonne, and Rebecca were there and Virginia served them tea.

All of life fits together

Pastor Simon looked at Moses and began speaking. "Moses, we Africans know that everything we do affects everything else in our lives. That is how God has made life. What we believe affects what we do. How we use our land affects the amount of food we get from the land. The amount of food we get affects our families. And how we raise our families affects the food, the land, and everything else.

"This morning you said the town of Malenga is growing. You have more and more people, but the land you have is the same. So you asked about how to get enough land and food for all the people of Malenga on the same amount of land."

"Yes, Pastor, that's right. As chief, I do not know what to do."

"Moses, let's talk first about God. Before God made the heavens and the earth, I believe he did a lot of planning. He thought about what he wanted to create. He decided how big the earth would be and the many kinds of plants and animals he would make. He then thought about how he would make us men and women to work with him. God is a God who plans.

"Remember that God has made us like him. That means we are like God who can plan ahead. If we want our lives to be good, we must plan our lives and our families. Only then can we have the abundant life Jesus wants us to have."

"Are you talking about family planning?" Moses asked. "I don't like to think about that. People who talk like that say we should stop having babies. Those people aren't helping us at all."

A family that has done no planning.

Planning for our family

Pastor Simon and the others laughed. "Moses," the pastor said. "Planning our families is much more than just about the number of children we have. It means thinking carefully about all aspects of the life of our family. Virginia and I talked about how many children we could raise well. We prayed and asked God for wisdom for our family."

Moses was astonished. “Pastor, we African men never talk with our wives about sex or how many children we should have. We just have them. God gives them to us, so why should we pray about it?”

The whole group laughed, and Robert spoke up. “Moses, that chair you are sitting on is a good chair. It holds you up well. If we put three big men on that chair, what would happen to it?”

“It would break.”

“It surely would,” Robert agreed. “The man who made that chair made it for just one person. The legs and seat are good for one person, but not for three.

How many children can live comfortably in your family?

“That chair is like my family. My wife and I are strong. I have a pretty good salary as a school director. But I am not rich. I know that my salary and our health and strength are strong enough for us to raise four children well. But if we had ten children, our family would be very unhealthy and our children would be unhappy. So we must talk about this and ask God for wisdom as to how big a family we should have."

Pastor Simon smiled, and added, "I knew that, as a pastor, I would not make much money. We also know that it takes a lot of money to raise each child. So we decided we would have only three children. They are now grown up and doing well. We have even been able to support our nephew John in his studies. God has helped us because we planned our family with him."

Virginia smiled and spoke up. "We also helped our children plan for their lives. We talked with them a lot about what God wants each of them to do. Our oldest son Paul is now studying to be a pastor. He wants to be like his father." She smiled proudly. "And he is doing very well.

"Our daughter Katherine is studying to be a nurse." At that point Yvonne clapped her hands and smiled. "And our youngest son wants to be a teacher. "It was now Robert's tum to smile and clap.

"Who will take care of your fields when you are old?" Moses asked.

Pastor Simon replied, "We are not sure as yet. If Paul decides to come here as pastor, he will care for the fields with his family. If another man comes here as pastor, he may want to buy our fields. We will leave that in God's hands."

David spoke up and asked, "Moses, how many children do you have?"

Moses said, "I have eleven."

Rebecca frowned and said, "Wow, Moses, your wife must have suffered a lot!"

Moses frowned. "Maybe, but that's what God said women would do."

Yvonne asked, "How old are they?"

Moses replied, "The oldest is seventeen, He has just started secondary school. The youngest was born last year."

David spoke up. "Moses, God also said that we fathers are responsible for our families. So what are all eleven of your children going to do?"

"I am not sure, "Moses replied. "We have not talked about that."

If we have many children, what can we do?

Pastor Simon spoke again. "Moses, in the past our families wanted many children to help grow food on the farm or raise cattle. But now farms are very small and cannot support many children. For each child we now need much money to buy clothes for them, to send them to school, and to buy medicines when they are sick. That is why it is important for us Africans to plan our family. This is sometimes hard for us Africans to do, yet God wants us to do that. He has planned for all of human life until the very end. We need to do the same with our lives. We should help our children do the same."

"What do you mean?" Moses asked.

"Talk with each of your children about their future. What would they like to do? We are going to help you see how you can make the soil of your fields fertile. It will then grow more food. If you can grow more food than you need for your family, you can sell crops like millet, groundnuts, and cassava to make money. Perhaps one of your sons will see that he can make a good living by staying in Malenga and farming like you do."

Robert added, "Talk to your other children about what we discussed this morning. Some may want to go to secondary school. Others may want to learn a trade. They can then learn how to make a good living, even right in Malenga."

Virginia said, "You see, Moses, God has given us good minds. God wants us to use our minds to think about our lives. God can give us his wisdom as we think about our lives if we ask him for it. We need to look at our whole life and see how everything fits together. We also need to think about the years ahead. We need to help each of our children plan for their future."

Moses scratched his head. "That is hard to do," he said.

Pastor Simon spoke up. "Moses, you remember the story of Joseph in the Old Testament. God gave a dream to the king of Egypt about seven fat cows and seven thin cows that ate up the fat cows. The king did not know what the dream meant. He heard that Joseph, a Hebrew young man, could interpret dreams. So he sent for Joseph. Joseph prayed and asked God for wisdom. God showed Joseph what the dream meant. Robert, you tell the rest of the story."

Robert continued the story. "Joseph explained that seven good years were coming. The fields would produce abundant crops. But then seven years of famine would come when nothing would grow. To save his people, the king should build big barns to store much grain during the next seven years. That grain would feed the people during the seven years of famine. The king was very impressed with Joseph's wisdom. He put him in charge of his whole kingdom to build barns and store grain. Joseph did that and saved the people of Egypt. He also saved his own family, the sons of Jacob, or Israel. By good planning, all the people were saved."

Moses shook his head and frowned. "Yes, I know that story, but I don't have the wisdom of Joseph."

The group laughed, and David said, "Moses, where did Joseph get his wisdom?"

Moses replied, "I guess he got it from God."

"Can't you get wisdom from God as well?" David asked.

How can we get wisdom from God?

Moses sighed deeply, then said, "Yes, I suppose I can, but I need help. And so do my people."

The pastor smiled and said, "Moses, God is always ready and willing to give us wisdom if we ask him." He read from the book of James:

> *If any of you lacks wisdom, he should pray to God, who will give it to him; because God gives generously and graciously to all. James 1:5* Good News Bible

Robert said, "Moses, you can pray and ask God for wisdom. You also have a Bible. The Bible gives us much wisdom about many things, even about our land and how to care for it. The Bible also teaches us much about how to care for our families and our children. The Bible even talks a lot about money and how to manage it well."

Planning for the community

Pastor Simon added, "Moses, you are the chief of Malenga. You should study the whole town and think about these things:

- *How many families are there in Malenga?*
- *How much land does our town have?*
- *How many acres (hectares)?*
- *How many acres are there for each family?*
- *Is there enough land to feed all the people?*

"Once you know these things, sit down with your leaders. Look at this information. Discuss together what to do so that your land will produce enough food for everyone. Discuss also what the children of all your families can do. Tell them what we talked about here this morning. Explain to them different ways to make money and have a good life. Help them see that this is God's way. God plans, and so should we.

"We will be glad to help you. We are neighbors, and God has told us we must love our neighbors as ourselves, and help them in every way we can."

Moses smiled and said, "Pastor Simon, I can see that you and your people here in Kalindi are God's people. He has helped you a lot. He has also shown you how to be kind to others as well. We will appreciate all the help you can give us."

Moses then asked another question. "Pastor, you told me the other day that, here in Kalindi, the roots of trees you had cut down were still alive. You let some of them grow and become trees again.

However, in our fields we cut down our trees many years ago. I think the old roots have died. What can we do?"

Pastor Simon agreed that most likely the roots of those old trees were no longer alive. So new trees could not grow from them. He suggested they discuss that problem during the evening after dinner.

The group spent a short time in prayer, thanking God for helping them talk these things over together and for putting wise thoughts into their minds.

Questions for study

1. *Jesus said, "If one of you is planning to build a big farm building, you sit down first and figure out what it will cost to see if you have enough money to finish the job. If you don't, you will not have enough money to finish the building after laying the foundation. All who see what happened will make fun of you." Luke 14: 28, 29 How does this teaching of Jesus apply to building our families?*

2. *When the population grows but the land remains the same, what must we do?*

3. *How can God help us think about the coming years and plan our families and our land?*

4. *How can we plan for the whole community so that life can be more abundant?*

Chapter 14

Moringa, the Miracle Tree

The women were having their monthly meeting at the home of Pastor Simon. Virginia, Simon's wife, served tea to the ladies. She placed a plate of crackers on the serving table and also a bowl of paste to spread on the crackers. Yvonne took some of the paste, spread it on a cracker, and began eating.

"Virginia," she said. "This is a delicious paste. Can you tell us about it?"

Virginia smiled and said to all the ladies, "Please help yourselves to the crackers and paste. While you are eating them and drinking your tea, I will tell you about a wonderful new tree we are growing here in our yard."

The 'miracle tree'

When the women were seated and munching on the crackers, Virginia began speaking. "Two years ago my husband went to a church conference in the capital. During the conference he learned about a wonderful new tree called the 'Miracle Tree.' Its real name is the 'moringa' tree. He brought home some seeds and planted them here in our yard. It is those trees you see over there."

"Why do you call it the miracle tree?" one of the ladies asked.

Virginia replied, "It is because the leaves of the tree have much protein in them. They also have vitamins and minerals that make us healthy. When you eat moringa leaves, you get much protein. When you feed it regularly to children who are poorly nourished, they quickly become much more healthy. It seems like a miracle when you see them getting stronger in just a few days."

Eating moringa tree leaves

"You eat the leaves of those trees?" Yvonne asked.

"Yes," Virginia replied. "We eat them nearly every day."

"How do you prepare them?" Yvonne asked.

Virginia smiled and said, "You are eating moringa leaves right now."

"We are?" the ladies exclaimed.

"Yes. I put the powder of moringa leaves, along with groundnut butter, in the paste you have spread on your crackers. If you eat a lot of crackers with that paste, my goodness, you will feel much more strong and healthy."

The women laughed and exclaimed they had never heard of such a thing. They wanted to know more about it. So Virginia explained to them how to eat the leaves of this wonderful tree.

Preparing moringa leaves to eat

"First of all," she said, "You pick the new tender green leaves at the ends of the branches. Come on over here and I will show you." She took them over to where the trees were growing.

She began pulling off tender leaves from the ends of some of the low branches.

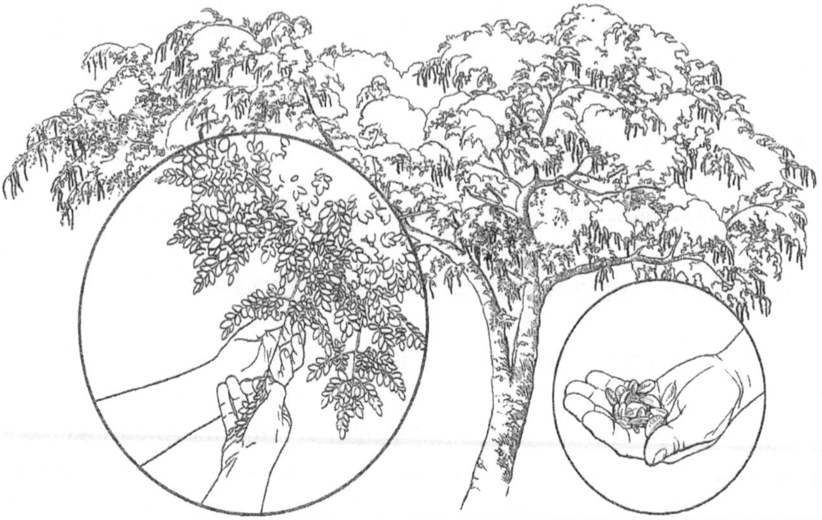

A moringa tree with seed pods. Pulling leaves off a moringa branch. Handful of moringa leaves.

"Collect a small basket of the leaves. Lay them out on a cloth or table to dry. Dry them in the shade, not in direct sunlight. If you put them in direct sunlight, the leaves will lose some of their vitamins. So dry them in the shade. It takes one or two days for the leaves to become dry and brittle: one day in the dry season, but two or even three days in the humid wet season."

"Then what do you do?" one of the women asked.

"When they are dry, you put them into the mortar and pound them into powder. Pick out the tough stems, and then pour the powder through a strainer or a fine wire screen. It is the powder from the leaves that you eat.

"You can mix the powder directly with the flour of maize, millet, sorghum or cassava. By doing that, this adds a wonderful protein to the staple maize, millet, or other paste that you eat. Or else you can add the powdered leaves to the vegetable or meat sauce you serve with the staple paste. Either way you will get much protein. As you can see, I mixed powdered moringa leaves with groundnut butter to make the paste you spread on the crackers. You could also use cream cheese instead of groundnut butter, if you can find it."

Moringa leaves - a complete protein

Yvonne spoke up. "That sounds interesting. But protein from green leaves and cereals, and even from legumes like beans and groundnuts, does not have a complete protein. Is it necessary to add another plant source of protein to moringa leaf powder?"

"Actually, no, it is not necessary to do that," Virginia replied. "The moringa leaf has all of the nine amino acids necessary to make a complete protein. That is why we call it the 'miracle tree.' Moringa leaves are as nutritious as meat, eggs, or milk and very easy to digest. They give a complete protein. It is not necessary to mix other sources of vegetal protein with the powdered leaves.

"Moringa leaves also have lots of vitamins, especially vitamins A and C. They have minerals as well, like iron and calcium to make your blood and bones strong."

Moringa leaves - safe and nutritious

Mary, a young woman, spoke up. "As you can see, I am pregnant. Can I eat moringa leaves?"

"Yes indeed," Virginia replied. "A nurse explained this at that pastors' conference. She said it is excellent protein for pregnant mothers and also for little children. It is perfectly safe. The leaves have nothing in them that can harm you or the baby. They also have lots of vitamins and minerals that make you strong. If you eat up to 6 tablespoons full of leaf powder a day, this will make you strong. Two tablespoons full of leaf powder every day will also help your baby to grow well."

She continued. "Here is another way you can serve the leaves. Take two cups of fresh leaves and steam them in one cup of water for a few minutes. Add a bit of onion and salt. This is like spinach and tastes real good. It has many vitamins and minerals that can make you healthy. You can feed this directly to young children."

Louise asked, "My youngest child is just a year old. Can she eat moringa leaves?"

"Yes," Virginia replied. "You can boil the leaves like I just said. Or else you can make a porridge of moringa leaf powder with the flour of maize, millet, or even cassava. This will now be rich in protein, excellent food for your daughter to eat. Three tablespoons full of moringa leaf powder every day will help your daughter grow to be healthy and strong."

"That sounds wonderful," Alice said. "I have seven grandchildren, and we are quite poor. As I understand it, the moringa leaves are a free source of good protein. They don't cost anything!"

"Exactly," Virginia replied. "You can also eat the seed pods when they are young and tender. Cut them into short lengths and boil them in water for ten minutes. Add onion and salt. They taste very good, and are good for you. You can also cook them with rice. That makes a delicious dish. Or you can roast the pods or fry them in oil."

She went on. "You can also make tea out of moringa flowers. Boil them in water for a few minutes. Add sugar, and drink it like tea."

The women discussed this for a few moments. They were impressed by knowing that a tree that already grows well there could provide them with so much food.

Virginia went on. "All of you can grow moringa trees and have a wonderful source of protein right in your own yards. If you have ten Moringa trees, you will always have enough leaves to give plenty of protein to your whole family."

"Tell us how to grow the trees," several of the women said.

Growing Moringa trees

"It is really quite easy," Virginia replied. "You need to know how to grow them correctly and keep them producing leaves." She led them to a tree that had many seed pods on it, and she pulled off some of the dry pods to show them. She split open a pod and showed them the dry seeds inside.

"You can plant these seeds in a nursery," she said. "You can also plant them directly in the ground where you want them to grow. That is a good idea, because then it is not necessary to transplant them."

She showed them a row of moringa trees growing along one side of their yard. Fifteen trees were growing in this row, and they were about 6 feet (two meters) apart.

"First, with your hoe make the soil loose where you want to plant the seeds. They should be about 6 feet (two meters) apart. Plant three or four seeds in each place. Put the seeds in small holes about 2 inches (5 cm) deep and cover them well. You can plant them more closely if you want to make them a living fence to keep out animals.

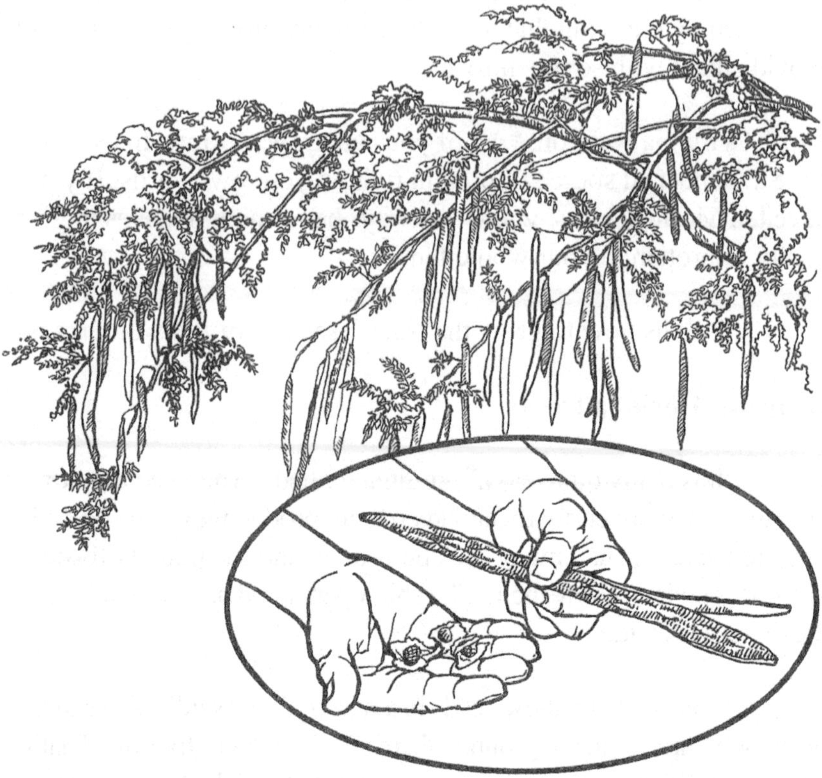

Branch of a moringa tree with many seed pods. Showing pod and seeds.

"If it is rainy season, or if you put water on the seeds regularly, it takes one or two weeks for the seeds to sprout. If three or four seeds sprout in the same place, remove all but one of them. Select the one that looks very healthy and strong. That way the one tree can grow well. My husband is here, and he can tell you how to help the trees to grow well."

Growing moringa trees by cuttings

Pastor Simon joined the group. He had just eaten a few crackers with the moringa/groundnut paste, and he asked the ladies if they had enjoyed the paste. They smiled and said, "Yes, very much. Now we are learning how to grow these trees in our own yards."

"You can also grow moringa trees by cuttings," the pastor said. "To do this, dig a hole 3 feet deep (1meter) and 3 feet wide (1 meter). Put good soil into it along with good compost or manure. Pack it down quite well and put a lot of water in it.

"Cut a one-year-old healthy branch from a moringa tree. Push the cut end down into the soil in the hole. Mound up the soil a little around the cutting. Keep watering it twice a week, unless it rains. Do this until the cutting is growing well and producing branches."

Trimming the trees

"The trees grow very fast and tall," the pastor said. "However, you do not want most of them to grow very tall. If they do, you cannot reach the leaves. So when they are as tall as you are, cut off the top of each one. You will then see that very soon, branches will grow out from the sides of the trunk. That is good, because it is those branches that will give you the leaves.

"As the trees continue to grow tall, keep cutting off the tops. The trees should stay only about six feet (two meters) tall. That way they will have lots of branches. However, let one or two trees grow very tall. They will be the trees that will produce lots of seed pods for you."

He added, "Cut off the ends of the branches regularly. They will quickly grow again, and in this way they will keep on producing tender leaves for you to eat."

He continued, "The moringa tree can grow in many different places. The trees grow well where it is warm and sunny. They can grow where it is quite dry and also where there is a lot of rain. However, if the soil is very wet, or even like a swamp, they will not grow well. They like soil that is full of sand, but they will also grow in soil that has a lot of clay as long as it does not stay too wet. So they can grow in many places.

"If the dry season is long, like 6 months, the leaves may fall off. You will then have no leaves to eat. In that case, put water on the trees frequently. That will help them stay green and produce leaves all the time."

Two kinds of moringa trees

"Pastor," one of the women asked. "My family lives up in the mountains. Will moringa trees grow well there?"

"There are two kinds of moringa tree. The kind you see here in our yard is called Moringa oleifera. It grows well all over this area. However, in the mountains above a certain altitude, it does not grow well. The other kind, Moringa stenopetala, grows better up there. It is much like the trees here and is better adapted to areas of high altitude. Your family can easily get seeds for that kind from the forestry service, or from an agricultural center or store."

"Are the leaves just as good as these?"

"Oh yes," the pastor replied. "You grow that kind of tree just like you grow the kind we have here. You use the leaves of it in the same way as well. The leaves are quite big. Those of Moringa oleifera are a bit smaller."

Anna spoke up. “Pastor, as you know, David and I live on a farm. We have goats and sheep, and also raise rabbits. Are the leaves good for those animals?”

“Yes indeed they are,” the pastor replied. “If you grow a lot of moringa trees, you can feed the leaves or branches to your animals and still have enough to feed yourselves.”

God’s concern for our food

The women continued their meeting. They were studying how God has given us plants and trees for food. Virginia read a verse from Psalm 104: 14; ‘He makes grass grow for the cattle, and plants for man to cultivate – bringing forth food from the earth.’

They turned to the pastor and asked, “Did our ancestors have this tree?”

“Yes,” he replied. “This tree has always grown here, and our ancestors knew about it. However, they did not know that the leaves were full of protein and good to eat. So they did not make use of it like we are doing now.”

“Isn’t it wonderful what we have been able to discover?” Yvonne exclaimed. “The knowledge and intelligence God has given us help us in so many ways.”

Virginia spoke up. “Here is a little book my husband got at the conference. It gives more details about growing moringa trees. Here is another one about how to use the leaves of moringa trees. It has many recipes using moringa leaf powder. If we use it in different foods, our children will eat it well.”*

As the meeting ended, they prayed to thank God for all he gives to us to make our lives good. They also thanked him for the intelligence he has given us to help us find out better ways to live and how to make use of the plants and trees he provides for us.

Questions for study

1. *What are the rules to follow to grow moringa trees where you are so they will give you good leaves to eat?*

2. *What are the different ways you can prepare moringa leaves to eat?*

3. *How do moringa leaves help children or adults who have poor nutrition to become healthy and strong?*

4. *How can the church help spread the knowledge of this 'miracle tree' to the people of your community and other communities as well?*

5. *In Genesis 3 we see that after Adam and Eve disobeyed God, growing food became difficult. Why is this so?*

6. *In Leviticus 26 : 1 - 13 we see how God can help us grow enough food. Study this passage carefully. How does it apply to us today?*

* Good books about the moringa tree and how to use it are available from the forestry service or from agricultural schools and centers.

Chapter 15

Growing New Trees for the Forest

The day arrived for the meeting about growing trees in order to replant the forest. A heavy rain fell during the night, and was still coming down slowly at dawn as Pastor Simon and Virginia ate breakfast. The pastor remarked how glad he was for the lessons they had learned about planting trees and doing other things to stop erosion. Robert, John and some of the others joined them. They drank tea together and waited for those coming from other communities to arrive for the meeting.

The benefits of the forest

By 10 o'clock a large group had gathered in the church building. They moved the wooden benches so they could all sit in a circle. Pastor Simon welcomed them and opened the meeting with prayer. Then he said, "Wonderful forests used to cover much of our land. They provided many benefits for our forefathers. Young people, and especially those who have grown up in the cities, have almost forgotten how important the forest is to us.

"The forest gave us much food – many animals, lots of fruits and nuts, and many small plants like mushrooms and herbs. We rarely see these any more.

"Forests were beautiful places to go – cool, much shade, magnificent trees, and many ways to see God's majestic works of creation. When I was a small boy, I would spend hours in the forest. It was a wonderful place to pray and talk with God. I would remember the words in Isaiah where God said:

'You will go out in joy and be led forth in peace. The mountains and hills will burst into song before you, and all the trees of the field will clap their hands. " Isaiah 55: 12

"But now the mountains and hills are bare. No trees are there to clap their hands. It is sad."

Robert spoke up. "The forests were a source of wealth for our forefathers. Poles for building houses, lumber for making furniture and other things, vines of all sorts for making baskets, ropes, and cords to

Some of the benefits of the forest; food, firewood, and building supplies.

make walls of our houses and fences around our yards, all these came from the forest. The big hardwood trees were very valuable for wood, but they are all gone now."

"Yes, Robert, that's right," the pastor said. "What's more, the forest protected our hills. Now you see many ravines that rainwater has made. This erosion is destroying the beautiful hills, and they seem to be weeping. This is why we now have many floods in the valleys. In past years, floods were rare. Now we have them because the rain falling on the mountains and hills just runs down those ravines into the valleys and causes floods."

Robert added, "My grandfather used to tell me that the forest gave us good weather. Somehow he knew that the vast number of trees kept the air clear and gave us good rain. Now our air is often heavy, full of smoke, and hard to breathe. He also knew that our springs of water were in the forest or on the edge of the forest. Those springs had good water because of the trees. But now many springs have dried up because the trees are all gone."

The results of cutting down the forest

"We have cut down most of the forest, "Pastor Simon said, "to sell lumber, to make charcoal, and to provide more fields for growing food. The results, however, have been bad for us."

"What bad results have come from this?" Moses asked.

The pastor replied. "Birds and animals are disappearing. Many springs of water have diminished or dried up altogether. It is hard to find firewood, or poles for building houses and fences. Lumber is scarce and very expensive. As Robert said, the air is now hard to breathe. The temperature is often very hot, and we do not have much rain now, less than I remember as a small boy."

Moses said, "We cut down much of our forest to make fields to grow food. We thought the forest soil was good. Isn't that true?" he asked.

"That's what we thought too," Pastor Simon said. "We believed the soil in the forest was good and would grow lots of food. Actually the forest soil is not good, for it is not deep. The minerals and compost are quickly taken up by the trees and vines. When we cut them all down, the sun quickly burns the little compost in the soil or used by the crops. In two years it is gone and we can no longer grow food on land where the forest once stood. The forest soil is fragile and we must take care of it."

The forest and the rain

Robert joined in. "Forests play an important role in bringing rain to us. Much water goes up into the sky from the forests to make rain. The water evaporates from the leaves. The forests absorb heat from the sun and make the air over them hot. This hot air draws wind from the oceans and lakes that also has much water vapor to make rain. When the forest is cut down, less water vapor is in the sky to make rain. The rains diminish, and droughts often occur. We must have forest to have regular rains."

Restoring the forest

Pastor Simon continued. "When you walk through areal forest, you see many different kinds of trees growing together. You do not find simply one type of tree. In many projects of reforestation, people plant just one type of tree, like acacia or pine trees. This is

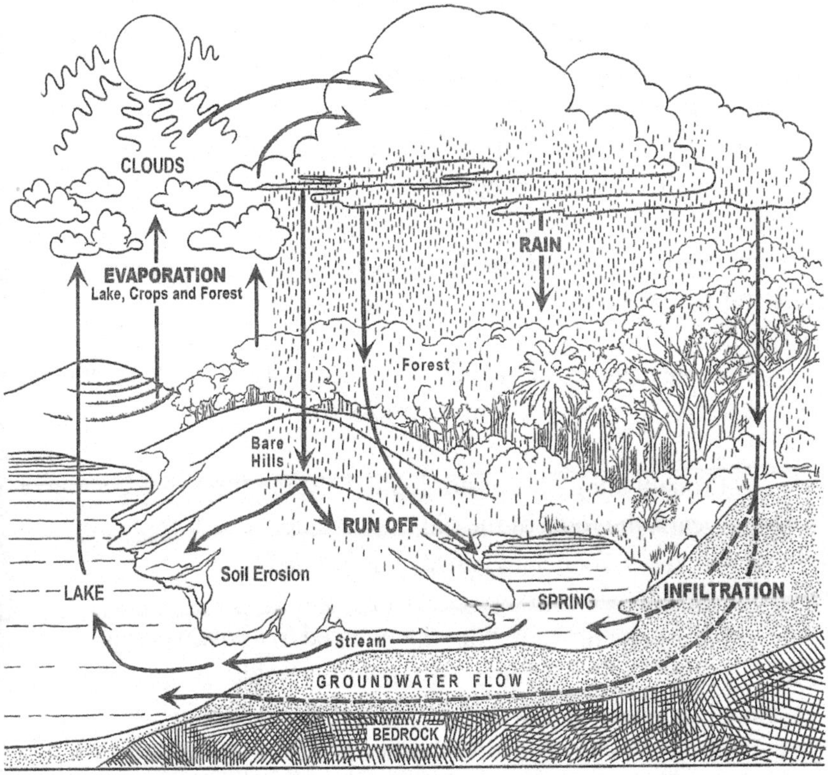

The forest enables rain to soak into the ground and renew springs and lakes.

useful for specific purposes, like producing firewood or charcoal, or for making paper. But this is not good for replanting the forest.

"As we try to restore the forest, we want many different kinds of trees to grow there. When different kinds of trees grow together, birds and animals will return. Mushrooms will grow, and we will have different types of trees for many purposes."

"How can you plant new forest trees?" Moses asked. "Where do you get seeds for that?"

The pastor replied, "If you cut down the forest just a few years ago, the roots of many of the trees will still be alive. If that is true, tell your people that no one should cut anything, make fields, or burn in the area where the forest had been. You will soon see trees beginning to grow again. After ten years, you will have a new young forest that will some day become like the forest of the time of your ancestors."

Planting forest trees

Thomas, the chief of Nakama community, said, "We cut down our forest many years ago. Will it ever regrow?"

"Probably not," the pastor replied. "If the forest has been gone for many years, the roots of the old trees are now dead. You will have to plant new trees. John, can you talk about this?"

John spoke up. "First of all, you need to know what kind of trees will grow where you are. Some of you older people may remember. In any case, it would be good to ask someone from the forestry service or our agricultural school to come to Nakama and see your situation. It would help to test the soil, look at the hills and flat places, and decide together what would be best to plant.

"You can get seeds of many different kinds of trees from our agricultural school. The forestry service can also help you. You may be able to get seeds of some trees from stores that sell agricultural materials. Or else, you can go into a forest and find seeds from the different kinds of trees that are there."

"How can you collect seeds from trees in the forest?" Moses asked.

John said, "Some trees produce a lot of seeds and they are easy to collect. First, clear the ground under a tree. Look there for seeds that have just fallen and still are healthy. You can then try to shake the tree, or hit the trunk hard, to see if more seeds will fall. If you can reach branches with seeds on them, you can collect those seeds."

"How do I plant them?" Moses asked.

"Plant them in your nursery in bags, or directly into seedbeds," John responded. "This would be a good project for the whole community. Talk to your people about the great need to restore some of the traditional forest. Get them to start a community nursery where seeds of forest trees can be planted. Then ask for an expert to come and give you advice about what trees to plant."

Robert spoke up. "You can also find new little trees growing up where there is forest. Dig these up carefully. Be sure to protect the roots from damage. Transplant them to your nursery and let them grow until they are three feet (a meter) or more tall."

"Can you cut branches and plant them?" Moses asked.

"Oh yes," John replied. "You can cut good branches from forest trees. They should be 2 to 4 inches (5 to 10 cm) thick. Be sure you plant them right away in your nursery. If you don't, the water in the cut branches will dry up quickly and the branch will soon die. So put water on them frequently. After some months, when many of the cuttings are growing well, you can transplant them to where you want them to grow."

Pastor Simon said, "It is good to plant the kind of trees that grew in your old forest. These trees are adapted to that area. They should grow well. If you try to plant trees from other areas, or even other countries, this may have bad results. The new trees may replace local trees and even damage the environment."

John added. "Remember that the government has rules about planting trees. Find out from the forestry service how we can help with this reforestation. They can tell us what kinds of trees to plant, where to plant them, and how to go about it."

Pastor Simon got up and looked out over the land and at the hills in the distance. "My hope," he said, "is that every family will plant a few new trees every year – fruit trees, forest trees, or trees to make the fields grow more food. If every family and every community did this, we soon would see our land being restored."

"Where can we plant forest trees?" Moses asked.

Where to replant the forest

The pastor answered, "You do not want the forest to grow back in the fields you need to grow food. But some of your land is not good for growing food. Where the hills are quite steep, that is a good place to begin planting forest trees. Or plant them where many rocks are in the soil and you cannot make fields."

Robert said, "On the sides of steep hills are ravines. You can see a few of them over there." He pointed to the hills south of Katindi. "Some of them are where little streams of water come down. Others are ravines made by rain erosion. These are good places to plant forest trees."

John added, "In the old days, a lot of land was just grassland, or savannah. Trees and forest did not grow there. That is because the water is very deep down in the soil and trees cannot grow well there. It is hard to plant forest trees in those grasslands. However, with much effort, you can grow trees there that are good for firewood or charcoal."

The pastor said, "What John is saying is true. Look for those places where the forest used to grow. Those places are excellent for replanting the forest."

Planting new trees

"How can we go about it?" Moses asked.

John replied. "Forest trees grow slowly from seeds or cuttings. It may take many months for a forest tree to grow to be 3 feet (a meter) or more tall. They need to be that tall before transplanting them. That is so they will have good roots.

"When you are ready to transplant new trees of different kinds, find the place where you want to transplant them. It may be in a valley by a stream of water, or in a ravine on the hill, or in rocky ground. Clear the place of brush and weeds.

"Dig big holes – at least 2 feet (60 cm) deep and wide. Keep the topsoil and the deeper soil separate. Be sure there is room in the hole for the roots of the young trees to grow down and out. If you can find manure, mix that with the topsoil. The holes should be 15 to 30 feet (five to ten meters) apart so the trees can grow big.

"Put a lot of water on the young trees before you transplant them. If they are in bags, take the bags to where the holes are. If the young trees are growing directly in the soil of your nursery, dig them up one at a time. Try to keep the soil on the roots from falling off. Even if it does, don't worry about it. Simply take the young tree quickly to the hole dug for it.

Planting a new forest tree.

"Put the new tree into the hole. Be sure the roots are straight and have plenty of room to spread out. Put the topsoil with manure around the roots. Fill up the rest of the hole with the deeper soil. Press down on the soil all around the tree with your feet. Be sure the soil comes up only to where the roots begin on the stem of the young tree.

"The soil in the hole should be a little below the ground around it. In this way, rain water will collect and add more moisture to the roots of the young tree. Put a bucketful of water on the young tree after you plant it. When there is no rain, put a bucket of water two times a week on each new tree. Do this for several months until the tree is growing well."

A new forest tree with a fence to protect it from animals.

Pastor Simon added, "Keep the ground clear of weeds until the tree is growing well. If possible, pile up grass and dead leaves around the base of the tree to make compost. Also, protect the little trees from animals. If necessary, build a small fence around each one to keep the animals away.

"Let's think about it this way:
In every valley, a small forest.
On every steep slope, trees of the forest.
In every ravine, more trees of the forest.
On the edge of every barren or desert spot,
trees growing to make the desert once again
become green and productive. "

Who will benefit from replanting the forest?

"I see one problem," Moses said. "I am getting old. These trees grow slowly. It will be many years before they will give us benefits. I will never see those benefits."

Robert leaned forward and smiled. "Moses, I am going to tell you something you may not want to hear. But it is true. When you and the people in Malenga cut down the old forest, you were not cutting down your own forest. You were cutting down the forest of your grandchildren. You were robbing them of their forest!"

"That is absolutely true, Moses," Pastor Simon said. "So think about your grandchildren. They will see many benefits from these trees and the new forest. They will say, 'We had a wise grandfather. He thought about us and planted these trees so we could benefit from them. We can thank God for our Grandfather Moses.'"

Then he added, "One other thing is important. We want to do this in order to restore the land God has given us. So we need to talk to God about this. Ask God to help you see good places to plant forest trees. When you plant tree seeds in the nursery, pray and ask God to help them grow well. When you transplant them to the places where you want them to grow, pray there to ask God to protect them and help them grow. We can plant seeds and young trees, but only God can make them grow. So we must do all of this with prayer, asking for God's help. If we work hard with the help of God, the mountains and hills will sing once more, and there will be many trees to clap their hands."

Getting every community involved

Pastor Simon went on. "I hope that we can help people in many places start tree nurseries. We should encourage every local church to do this to help their people plant new trees. Every primary and secondary school should have a tree nursery. That way, the children and students can begin to learn the importance of fruit trees and forest trees and see how to take care of the land God has given us."

Robert said, "I will talk about this at the next meeting for school teachers and supervisors."

Thomas said, "All of this sounds good. I may even try to start a small nursery at my home. That way, I can teach my own children about it. We could grow fruit trees and have more fruit to eat. We can start acacia and leucaena trees to plant in and around our garden and fields. This will help our family have more food."

John added, "Try to get the whole community to start a community nursery to help all the people. In this way, they can plant forest trees on all their unused land. Soon there will be many forests growing, and the land will be better."

The pastor said, "Yes, then animals will return to the forest. Many birds will come back. And people can then use wood and trees from the forest for fuel, for building houses, and for making furniture and other things."

Moses said, "Thank you for telling me these things. But you keep mentioning something else and saying we must protect new trees from animals. That is a big problem in our community. Goats, chickens, pigs, and even cows run everywhere. They get into our gardens and eat our vegetables. They often break into our fields and eat our crops. What can we do to stop this?"

Thomas immediately spoke up. "We have that problem also, and we don't know what to do about it. Pastor Simon, can you help us?"

The pastor replied, "Ladies and gentlemen, this is an enormous problem and very difficult to solve. We Africans have always allowed our animals to go where they want to go and find their own food. But much of the food they find is our food. So until we learn how to control our animals, we will never have productive farms and sufficient food."

They all agreed, and asked, "So what do you suggest?"

Pastor Simon answered, "I suggest we have a big meeting to talk about this. Can we all gather a week from Saturday morning back here and talk about this?"

After some discussion, they all agreed to come back to Katindi to consider this difficult problem – how to control animals.

The pastor prayed, asking God for wisdom and courage to replant the forest and restore the marvelous forests he has given us. He said he would welcome them back on Saturday, and then reminded them of the other meeting in two weeks in the government center. They would meet there with government leaders to talk about how to grow trees that could provide fuel for them.

Questions for study

1. *In how many different ways does the forest help us?*
2. *In what ways are we suffering now because much of the forest is gone?*
3. *What can each family or community do about it?*
4. *Since God planted all the forests of the world before he created man, why should we plant forest trees now?*

5. *How can you, or the whole community, start a nursery of forest trees?*

6. *Where are places near your community where you can plant forest trees?*

7. *If you do all of this, what will the farms and hills around you look like twenty years from now?*

8. *Read Isaiah 55: 8 - 12. What happens when rainfalls abundantly from the sky? Why does God compare the rain to his Word? What happens when the Word of God comes into communities abundantly? Will that help us have more food? How?*

Chapter 16

The Government, the Church, and Trees

Pastor Simon looked at the envelope and frowned. It was an official letter from the office of the District Commissioner. "What does he want with me?" he wondered. He frowned even more deeply as he read the letter.

Dear Pastor Maluba,

It has come to our attention that you and your church are engaged in promoting agroforestry. I am writing to inform you that this is the domain of the government and not of the church. Your work is to encourage the religious and spiritual activities of your people. All social activities are the work of the government.

You are requested to come to a meeting next Wednesday morning at ten o'clock in the district Commissioner's office to explain your activities.

With sincere wishes,
Arthur Kilomono, District Commissioner

Pastor Simon spread the letter out on his table and began to pray. He talked to God about the problem and asked for wisdom as to how to proceed. He needed God's help as to how to explain to the government officials the deep concern he had for the land and the forest.

His wife Virginia came into his office and he showed her the letter. She suggested that he call some of the church leaders and ask for their help as to what to do.

That evening several church leaders met with Pastor Simon in his living room. He read the letter from the District Commissioner to them. Robert was the first to speak.

"Pastor, I think the government is concerned that we are doing agroforestry projects and have not told them anything about it. If indeed we do any project, we should inform the government. But we have not done a project. We have simply been talking to people to encourage them to take better care of the land."

Ruth agreed, and added, "I think some of us should go to the District Commissioner's meeting with Pastor Simon. He should not go alone." They all agreed and appointed Robert and Rebecca to go with him.

On Wednesday morning Pastor Simon, Robert, and Rebecca arrived at the District Commissioner's office. The pastor showed the letter to the secretary and explained that he and two of his church members had come to meet with the Commissioner. The secretary showed them into the meeting room. A few minutes later, the Commissioner with three of his deputies came into the room. They greeted the church delegation and shook hands with everyone.

The Commissioner began the meeting by thanking the pastor and his leaders for coming to the meeting. He read to them the responsibility of the government in agroforestry. He said that it was very important that the activities of the church and the government should be kept separate. He expected the full cooperation of the church in this matter. He then asked Pastor Simon to explain why the church had become involved in this.

Pastor Simon thanked the Commissioner for inviting them to this meeting. He said the church was grateful for the concern of the government for the land and the forest. He expressed appreciation for the efforts of the government to preserve the land and reforest certain areas.

"Mr. Commissioner," he went on. "We believe that the health of the land should concern everyone. We have been encouraging our members and also leaders in our communities to take better care of the land. We are suggesting that they plant trees that will be of benefit to their fields, gardens, and springs of water. The trees will make the land more fertile and preserve the springs of water."

Robert spoke up. "Mr. Commissioner, you are rightly concerned about the church doing projects without informing you of what we are doing. Actually, however, we have not engaged in any projects. Pastor Simon has been pointing out to us in the church, and also to leaders in our communities, that it is God who has given us the land and the forests. God has given us wonderful resources He has also given us the responsibility of taking good care of these resources. We realize we have not taken good care of them. So we have a responsibility to God and also to the government, to ourselves, and to our grandchildren to take better care of these resources."

Pastor Simon added, "Mr. Commissioner, we in the church are very anxious to cooperate with you and the government in every way. Please tell us how we can be of help to you."

The Commissioner told the Pastor and the others about the projects of reforestation that the government was proposing. He also described the legislation about forbidding people from cutting down trees and the places where farmers could not make fields. "However," he said, "many people do not follow the rules we have made about the forest. They are not willing to cooperate with us in the reforestation projects we are proposing. They say the land belongs to them because their ancestors have given them that land. They do not want us planting trees on their land."

One of the Commissioner's deputies added, "In spite of our efforts to teach the people about the importance of trees, they do not seem to understand."

Pastor Simon replied, "Perhaps we can help you with this. You and we understand the importance of the forest and what is needed to keep the land healthy. We know the scientific explanations of this.

People in the communities, however, have not studied science. They do not understand the physical causes of problems very well. But they do know about God and that he is very important in our lives. So we are trying to show them what God has said about the land and the forest."

The Commissioner looked surprised. "What does God says about these things? We know only what science says about agriculture and forestry."

Pastor Simon opened his Bible and read to the Commissioner how God created all of nature. He planted trees, plants, and flowers over the whole earth. Then he created men and women to take care of them. God says that the whole world belongs to him. This includes the forests and valleys, the springs and the rivers, and all of nature. He has given to us these wonderful things to use and made us responsible to take care of them.

The Commissioner was astonished. "I never heard that before," he said. "Oh yes, I believe in God. But I assumed God simply was concerned about heaven and what we must do to get to heaven. I was not aware that God is concerned also about how we live here on earth."

"Oh indeed he is," the pastor replied. "In fact, Jesus tells us that the responsibilities we will have in eternity in heaven depend on how well we carry out our responsibilities here on earth. The Bible also tells us that it is God who created government. God knows we must organize our lives and rule over ourselves, not just by our own wisdom but with his wisdom and help."

The Commissioner leaned forward in his chair. "You mean God is concerned about how I rule over this district?"

"Absolutely," Pastor Simon replied and smiled. "I am sure God is pleased that you are trying to care for the land and the forest. It is God who has made us in the church concerned about these matters as well. That is why we are trying to help our people and communities understand the importance of working together to take better care of the land and the forests."

The Commissioner was silent for a moment. He then motioned to his three assistants and they excused themselves from the room. While they were gone, Pastor Simon, Robert, and Rebecca prayed quietly. After a few moments the Commissioner and his assistants returned. The Commissioner thanked the group for telling him things he did not know. Then he said, "It is clear to me that we all share the same concern about the land and the forest. I can likewise see that you have much wisdom. You are close to the people. We government leaders are not close to the people. It is therefore important that the churches and the government work together. So Pastor, I would like to invite you to be a member of the district committee on land and forests."

Pastor Simon was pleased to hear this. He replied that he would be happy to be on that committee. He could represent the churches in the area. He said he could also represent the government to the churches. He would explain to them how the government and the churches can work closely together.

The Commissioner turned to Robert and spoke to him. "You are the director of the school in Kalindi. I believe you also should be on this committee. You can then exercise your position in the school system to educate children and students better as well as the teachers about what they can do to improve the land and the forests."

Mr. Naninga, the Deputy Commissioner, spoke up. "We see very clearly here a spirit of cooperation. This is extremely important if we are to help our land develop and become prosperous. I believe this is an important day in the life of our district. We want you church leaders to help the people in the churches and in your communities understand that we, in the government, want to make our land better. We cannot do it alone. We also cannot do it simply by making laws and regulations. But if we can all work together, we can restore the land and build it up. We will then all become prosperous."

Pastor Simon responded by saying how happy this expression of cooperation made him. He assured the government leaders of the full support of his church and his efforts to help all the churches and communities work together with the government. They all stood and shook hands. The Commissioner asked Pastor Simon to close the meeting in prayer.

Before they left, the Commissioner said that one month from next Saturday, he was holding a meeting about growing trees for fuel. Because of the need of wood for fuel, people were cutting down much of the forest. He was concerned about finding ways to grow trees especially for fuel. A professor from the agricultural school would be there. He would like church leaders to come as well. Pastor Simon said he would be happy to come, and he would invite leaders of other churches and communities around Kalindi to attend as well.

Questions for study

1. *Read Romans 13: I - 7. What can we learn from Paul's instructions about the relationship between the government and the church?*

2. *Why is it important for churches and the government to work together?*

3. *What problems need to be overcome to improve this cooperation?*

4. *Can God work through government leaders to accomplish his purposes?*

5. *Why is it important that we pray for government leaders?*

Chapter 17

Protecting the Land from Animals

The following Saturday morning a group of leaders from several villages, including Moses from Malenga, gathered at the house of Pastor Simon. They sat together under a big mango tree and Pastor Simon's wife served them tea. Several of them remarked they had seen no goats or pigs wandering around the village of Katindi.

The problem of wandering animals

"Goats and pigs used to wander freely in our village," Pastor Simon said. "They ate peoples' gardens and any new bushes or trees we tried to plant. It was frustrating. At the same time many of our children had poor nutrition, but the goats and pigs were quite fat. We knew something was wrong."

Henry, the chief of Kimbata village, laughed and said, "You mean the animals were eating what was intended for the children."

"That's right," Pastor Simon said. "So we knew we had to do something about it."

Moses asked, "What did you do?"

Pastor Simon replied, "We realized this was a major problem. In fact, this problem has existed from the time of our ancestors. No one has found a real solution to it. But we knew the time had come to do something about it. We could not have the kind of life and prosperity we desire if we let animals wander freely and destroy crops and gardens. John, tell them what we talked about."

John explained, "We know that in Europe the people solved this problem many years ago. They have wonderful farms and grow many crops. They have other farms where they raise cows, sheep, goats, and pigs. They keep the farms for cultivation completely separate from animal farms. So they have much food, including meat, milk, and milk products."

Moses objected, "We have already seen that European methods don't work here in Africa."

"That is true," John replied. "However, the principle of keeping fields safe from animals is true everywhere. But we must find ways to do it that will work here in Africa. Africa will become prosperous only when we solve this problem."

Henry spoke up. "Pastor, we talked about this in our village, but our people objected. First of all, they said God did not make man to be a slave to animals. No way would they spend time seeing that animals got enough food and water. Animals should take care of themselves. The second problem was that this would cost money – money to pay someone to look after them, or to buy fencing to keep them in one place. We could not solve those problems."

What God says about ruling land and animals

"Henry, we discussed both of those problems," Pastor Simon said. "We talked about how men should live with animals. Animals are important. They give us food. They provide wealth for us. So they are valuable to us.

"Right at the very beginning God told Adam and Eve to rule over the animals. We find that in Genesis 1: 28 where it says, "*God blessed the man and the woman and said to them.... 'Rule over .... every living creature that moves on the ground.'*

The real question for us is: Who rules here? Do we rule our animals and our fields, or do our animals rule over us and the fields? Frankly, up to this time in Africa, it is the animals who rule. That is why we have so many problems growing food and becoming prosperous."

Which man rules over his animals?

"How are we supposed to rule over our animals?" Henry asked.

"It means keeping the animals where they belong while at the same time seeing they have what they need," Pastor Simon replied. "This is not being a slave to animals. Rather, it is keeping animals in their proper place. In that way we control them and they make money for us. God says, '*A righteous man cares for the needs of his animals* (Proverbs 12:10)'. We should do that."

"What did your people think about that?" asked Henry.

"Our people recognize that God has made a wonderful world and has shown us the best way to live in it. When they realized that God does want us to have animals and rule over them, they were able to accept this idea."

"But caring for animals and especially buying material for fences costs a lot of money," said another chief.

Caring for animals properly increases wealth

"Right, "said Pastor Simon. "But it can make even more money for us if we do it properly. So we talked about what it costs to raise animals properly. Animals are wealth. Think what you can get for a fully grown cow, or goat, or pig. A lot of money. When animals are confined in a small area and do not have to walk far to get their food, they grow fat quickly. You can sell them soon, and that increases your wealth. But if you let them wander in search of their own food, they grow slowly. You must wait until they are several years old before you can sell them."

"I think I see what you're getting at, Pastor," said Moses. "You have to spend some money in order to make more money."

"Moses, that's right. And our people began to see that."

"So what happened to the animals?" Henry asked.

"First of all, our people are beginning to think ahead. They realize that if we spend a little money and effort now, we will get more money next year. We can then spend some of the money we make next year to make even more money in the future. That is how we become prosperous and get ahead. This is what Jesus taught his followers to do." (See Matthew 25: 14-30)

Amos from Kilusu asked, "Pastor Simon, what did you do here in Katindi?"

We must respect the property of others, including animals

Pastor Simon replied, "Some of the leaders of the village wanted to kill all the animals. But we know we could not do that. The animals belong to people and we could not kill them."

Oscar, another chief said, "But those animals that belong to people eat the gardens and crops of other people. So why couldn't you kill them?"

"A good question," replied Pastor Simon. "We discussed that for quite a while. It all came down to respect for the property of other people. We saw this in the Bible. Even Moses told the people of Israel: '*I have not wronged any of them; I have not even taken one of their donkeys*' (Numbers 16:15). If he, the great leader of Israel, respected the property of the people, we should do the same."

"So how did you go about it?" another chief asked.

"We realized we had to educate our people about property and the rights of each person to his own property. Animals belong to people and we must respect that. So do the gardens and fields of people and the crops they have planted there. The problem was that the right to own animals was conflicting with the right to own gardens and fields. We had to find a solution that would respect both rights."

"That sounds pretty complicated to me," said Moses.

"Very complicated," agreed Pastor Simon. "Yet not impossible. We spent a lot of time discussing how to raise goats, cows, and even pigs in such a way that they would not destroy the property of other people. This meant keeping the animals in a separate place where they would have plenty of food and water. They would not be allowed to get out and wander freely and eat other peoples' food. We talked about different ways to do that, didn't we John?"

That's right, Uncle Simon. We discussed different ways to do this.

1. One way that costs very little is to tether animals. The animal is tied to a stake or a tree and can eat the grass in a circle around it. The owner then will move the stake to another location after a few hours. This is a good way to control animals. They can often eat grass in out-of-the-way places and comers.
2. Another way is to have fences around the fields so wandering animals could not get into them.
3. A third way is to put fences around the fields for animals (pastures) so farmers can raise animals to make money."

Living fences to keep animals in their place

Amos from Kitaka spoke up. "Fences cost money. We can't afford to buy fences for all our fields."

The pastor replied, "You don't have to buy fences, Amos. Around most of our fields here in Katindi, we have living fences."

"What in the world is a living fence?"

Moses smiled and said, "Pastor, I have seen your living fences. Can I describe them?"

"Sure, Moses. Go ahead."

Moses explained how Pastor Simon and his neighbors have planted fast-growing trees on the borders around their fields. Around some of the fields they have planted leucaena trees. Around others they planted Acacia mellifera trees. Around big fields they even planted grevillea trees. These trees have lots of seeds, so they are easy to plant. They are about 3 feet (one meter) apart, and it is easy to cut palm fronds, or bamboo stalks, and tie them between the trees to make a solid fence.

Pastor Simon continued. "We do not let the trees grow very tall, or grow out over the fields. We do, however, let the branches grow together. We cut off many branches and lay them between the growing plants to cover the soil. That keeps the moisture in the soil and protects it from the sun. The leaves of the branches decompose and become compost. And the tight fences keep the animals out, even the chickens."

"How do you get into the fields?" Amos asked.

"Oh, we leave a space and put up a gate. It is very simple. Later we will walk out and show this to you."

John added, "You can also use other plants to make a living fence. Cactus plants, like sisal, that have sharp points on their leaves, make a good fence. You plant them close together, and animals can never get through. Bushes with many thorns can also make a good fence."

"But then you can't move your fields from year to year," Henry objected.

"By doing it the way we do," Pastor Simon replied, "you do not need to move your fields. You can plant them nearly every year. The leucaena and acacia trees add nitrogen to the soil. The branches we cut protect the soil and add compost to it. We also plant different crops in each field every year, and that helps keep the soil strong."

Animals are a community problem

Pastor Simon continued. "We must realize that the problem of animals wandering and destroying things is a problem of the whole community. Each community needs to make rules. They must make sure the people follow the rules. For example, if you have a garden and my goat gets into your garden and destroys much of it, the community decides how much I must pay you for the damage. In that way, I do my best to keep my animals tethered or fenced in so they will not destroy the fields or gardens of other people."

Raising animals on farms

Oscar said, "Pastor, you mentioned that some farmers just raise animals. They do it in separate pastures with fences around them. Tell us about that."

"The best place to raise cows, goats, or sheep is in a grassy valley where there is a stream of water or a pond. A farmer can plant fast-growing trees around a few hectares or acres of grassy land. By tying palm slats or small branches between the tree trunks, he makes a living fence. A couple of our cattle farmers prefer planting thorny cactuses, for they work well as fences. The grass makes good pasture, and the animals drink water from the stream or pond."

"Do they also raise crops?" Amos asked.

"Some of them do, "Pastor Simon replied, "but in separate fields also surrounded by living fences. They use the manure of the cattle to spread on the fields to help the crops grow better. Some of those farmers also sell some manure to farmers like me who do not raise animals. We put it on our fields or gardens to make them fertile. That way their animals make more money for them."

Henry said, "You mean some farmers grow crops and raise animals at the same time? Do any of your farmers just raise animals?"

"Yes," the pastor replied. "Three of our farmers only raise cows and goats. They work hard to do that and make quite a lot of money from them. They do not have time to grow crops. The others, however, raise both animals and crops in separate fields. Then others, like me, only raise crops. But we no longer have the problem of cows, goats, or sheep getting into our fields. They are all kept in fenced-in pastures of other farmers."

Henry said, "This sounds interesting. I wonder if we could do that in our community."

John spoke up. "We have experts at our agricultural school who would be glad to come to Kimbata and show some of your farmers how to do that. They could teach them about how to rotate pastures, how to keep the pastures properly, and the medicines the animals need."

"When did you begin separating fields from pastures?" another chief asked.

"We started this four years ago, and it has gone well," Pastor Simon replied. "Three farmers now have herds of goats and cows, and that is all they raise. Others have smaller herds and also fields for crops. The animals are growing faster, and are in good health. The farmers buy vaccines and tick medicine from the agricultural school to protect the animals. They are now selling cows and goats. They get a lot of money. They spend only a little the money. So they are happy."

John added, "The people see that animals that stay in a pasture grow big quite fast. Those that wander do not grow big or fast. They see the real advantages of keeping animals in a smaller area where they have food and water."

Moses asked, "Why is that so?"

John replied, "It is because the animals on these farms can use the food they eat to grow. So they grow big and fast. The animals that wander use most of the food they eat just to wander around looking for more food. So they stay quite thin."

The problem of many animals in small pastures

"How many cows can you raise in a pasture?" Moses asked.

"That's a very important question, "John answered. "It depends on the amount of grass available. It is not good to raise more cattle in a pasture than the grass can feed. If a pasture is fertile, the grass will be good. A hectare, or 2 to 3 acres, of good grass can support up to four cows or a dozen goats. If a farmer has more than four cows for each hectare or 3 acres, the pasture will soon become weak and the cows will not grow well. A cattle farmer needs help from someone at the agricultural center or school. A cattle expert can help a farmer know how many cows or goats he can raise on the land he has."

Pastor Simon spoke up. "Let me tell you an important story from the Bible, from Genesis chapter 13. Abraham and his nephew Lot both had large herds of cattle. Their herds kept growing. Soon they realized the land would not support so many cattle. The herds were eating grass fast, and the grass could not grow back fast. The land was becoming poor. Abraham was wise. He talked with Lot, explained the problem, and suggested they go to separate places with their herds. So Lot went one way, and Abraham went another way. That way there were fewer cattle and goats, and the grass and soil could become strong again."

Henry asked, "So what are you telling us, Pastor? Should we move our villages?"

"No, I don't think so. In Abraham's time, there was much land. Today, however, we have many communities and only a little land. However, we need the wisdom of Abraham to see what is happening to our land, our soil, and our pastures. John has told us how many cows can grow well on a certain area of pasture. We must work hard to keep those pastures strong. When we have more cows and not enough pasture, what pasture we do have will become weak. When that happens, we should sell some of the cows and keep only what the pasture can support."

John added, "That is where our school of agriculture can help. We have teachers who can examine pastures, see what the soil and grass are like, and tell you how many cows the pasture will support."

"Pastor Simon, tell us about the vegetable gardens here in Katindi and your fields." Moses asked. "Are they doing better?"

"Absolutely, "Pastor Simon replied. "Animals are no longer getting into the gardens or fields, and the food production is growing. Our children now look more healthy. I often buy buckets of manure from the cattle farmers to put on my garden. That helps me get more vegetables."

"What about pigs?" Oscar asked.

"Pigs are difficult," said Pastor Simon. "It is difficult to build a fence that keeps pigs inside it. They can dig holes under any fence. So we had to find another way for them."

"What did you do?"

"My nephew John discussed this with his teachers in the agricultural school and they showed him a good way. John, tell them about it."

John explained to the group what he had learned at school about raising pigs. He then took the group to Roger's house to see how Roger was keeping his pigs.

A farm to raise pigs

Roger had two big pigs, one male and one sow, down in the valley below the village. He built a pigpen out of strong posts and wide boards. The bottom board on each side went some inches below the ground to keep the pigs from digging under it.

Each pig was in a smaller pen inside the big pen. The sow already had a litter of pigs. Roger was building another pen next to the big pen because some of the piglets would have to go into the new one when they were weaned. He would sell them after one or two years when they had grown big.

Feeding the pigs took a lot of work and some money. Roger grew maize and peanuts, some of which he fed to the pigs. He also purchased a few bags of feed from an agricultural store in the city. All food garbage from his house went to the pigs. He gave them bananas, mangoes, other fruits, and some forage.

He explained how he kept the pigpen clean. Each morning he removed the manure and uneaten garbage from the pens. He kept that manure in a small pen nearby. He said that made good compost for his vegetable garden and for his field of maize. He said it is very important to keep the pens clean to protect the pigs from diseases. He also sells some manure to other farmers.

Below the pig pen was a small pond. Roger explained how he had dug that pond and was growing fish in it. Some of the manure from the pigs went into the pond. That helped the tiny plants to grow that the fish eat. So he can harvest quite a few fish from his pond.

Henry asked Roger a question. "You do a lot of hard work to raise these pigs. Why do you do that?"

Roger smiled and answered, "Because I will make a lot of money on these pigs. There are six piglets in that litter. Next year I will begin selling them. People pay good money for a grown male pig and more money for a female. The sow will have another litter next year. Each year I will have from four to eight pigs to sell. The money I get from selling them is much more than the money I am spending on food and medicine for them. So my hard work is making money for me and my family."

Pastor Simon added, "The way Roger is raising these pigs is good for the community. It is showing others how to do it. It is not harming the land, and the pigs are not destroying peoples' gardens or fields."

The men all shook Roger's hand. They were glad to see how well his pig project was doing.

Henry spoke again. "I have a brother who lives up north. They have a different problem up there. Large groups of people live there who have many cattle. They do not have farms to cultivate crops. They simply raise cattle. They are nomads and go with the cattle to wherever there is grass for forage. Often those cattle get into the fields of the farmers and destroy a lot of crops. Fights break out and somc people are wounded and even killed. They don't know what to do about it. Pastor, do you or John have any words of wisdom about that?"

"This is a major problem," the pastor replied, "because they have lived this way for thousands of years. A real change in culture has to occur, and that will take time.

The farmers and cattle herders must talk together. They need to decide what land belongs to the farmers and what land belongs to the cattle herders. They need to work out ways to respect the property of each other. The farmers need fences – wire fences or living fences – around their fields. The cattle herders must keep watch over their cattle to be sure they remain on the land that is theirs. It will take a long time to do this. We must care for the land wisely as God wants us to do. God also wants us to live in peace with those who do things in different ways.

"My brothers," the pastor continued. "Let's now go and see the living fences around our fields. We can then go down into the valley where the animals are being raised and talk to the cattle farmers there. Then we will come back and discuss any other questions you may have. And let me remind you about the meeting Saturday in the government center to discuss how to grow trees for firewood."

Questions for study

1. *List some of the problems that come when animals are allowed to go wherever they want to go.*

2. *How can we make living fences to keep animals in their own fields (pastures)?*

3. *Why do animals grow better when they are raised in a small area of land?*

4. *What must we do to raise animals well in such an area?*

5. *Why can farmers and cattle herders make more money if they keep fields for crops and pastures for cattle separate?*

Chapter 18

Trees for Fuel

In the conference hall of the district headquarters, a large group of community leaders had gathered. Pastor Simon, John, Robert, Yvonne, and Francis, the headman of Katindi were there. Community and church leaders from many towns in the district were there. The District Commissioner sat on the podium along with leaders from the agricultural school.

Cutting down the forest for fuel

After welcoming everyone, the Commissioner introduced Professor Kingwedi from the agricultural school. The Professor talked briefly about the necessity of trees to provide fuel for people, those in cities and towns, and farmers in the rural areas. He said that the need was great. Trees are the only source of fuel for most people. Because of that, many people were cutting down trees from the little forest that remained to make firewood and charcoal. Soon no trees would be left. He then asked for comments from the group.

Joel was the headman of Kingala, a community near to the district city. He asked a question of the group.

"We had a good forest years ago. However, people began cutting down the trees to burn the wood to make charcoal. As you know, charcoal is the main source of fuel in the city. Many merchants come to our community to buy sacks of charcoal. Now our trees are almost gone, and we have no trees left for ourselves for fuel or for building houses. What can we do?"

Using Eucalyptus trees for fuel

Amos, headman of Kitaka, spoke up. "Our community is also quite near the city. Some years ago, we learned how to grow eucalyptus trees. They grow fast, and within a few years they are big. They make fine poles for building houses and fences. They also make excellent firewood. Several of our farmers now grow only eucalyptus trees. They make a lot of money selling firewood for people in the city."

Henry, the headman from Kimbata, spoke up. "We also have been growing eucalyptus trees. It does indeed make good wood for fuel. However, these trees are hard on the soil. We can't grow anything under them, or even near them. The soil has become very poor. We want to find some other way to make firewood that will not make the land tired."

A plantation to grow trees for fuel

The professor now spoke to the whole group. "There is a better way," he said. "Last year, I took some of our students to see a tree project. It is out on the plateau near the town of Mampo and not far from the capital city. The project grows a species of acacia trees called Acacia auriculiformis. The wood of these trees makes fine firewood and charcoal."

"Tell us about it," the group said. "How do they go about it?"

"Each farmer has a large piece of land to grow these trees, about 8 hectares (20 acres). It takes eight years for a tree to grow big enough to cut for charcoal. The first year he prepares a plot of one hectare (2.5 acres) and plants acacia seeds he gets from the agricultural school. He plants them in rows about three meters (10 feet) apart. Between these rows he plants his regular crops like maize, millet, cassava, groundnuts, beans, sweet potatoes, and squash for himself and his family."

"That's a lot of land. How long does it take him to clear it all?" someone asked.

"He begins clearing the ground during the dry season. He can clear the plot of one hectare (2.5 acres) in a couple of months, especially if his whole family works together. He burns the big trees and bushes he has cut down. He leaves the weeds and grass he has pulled out on the ground to make mulch. He does not plow the ground. When the rains begin, he plants the acacia seeds and also his crops. This way the seeds can sprout and the young trees get a good start before the rains stop. They continue to grow during the next dry season."

Planting acacia seeds. One year old trees with crops planted in between the rows are in the background.

Amos asked, "How do you plant acacia seeds?"

The Professor replied, "You plant them in loose soil, well chopped by a hoe. With your finger make three holes close to each other, about 5 cm (two inches) deep. Plant a seed in each hole. If all three seeds sprout, pull out two of them so there will be just one tree in each spot. Be sure to keep weeds from growing right around the little tree. The farmers in Mampo do that weeding as they cultivate and weed their crops."

Henry said, "I suppose during the next dry season he cuts another hectare plot (2.5 acres) to plant trees there."

"That's right," the Professor said. "The second year he plants his crops again in the first plot between the small acacia trees. These little trees now produce some nitrogen for the soil, and his crops will grow better. In the meantime, the seeds sprout in the second plot and the young trees start to grow."

"I think I get the picture," Henry said. "Each year he clears another hectare plot (2.5 acres) and plants acacia seeds in it. In eight years all 8 hectares (20 acres) are planted in trees."

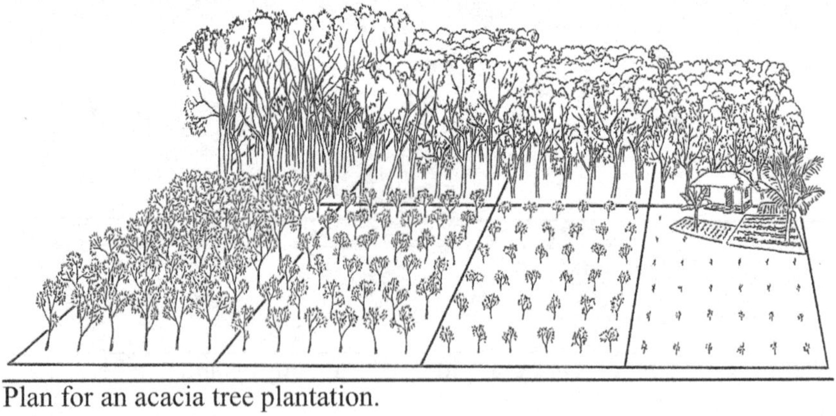

Plan for an acacia tree plantation.

"Exactly, "the Professor said. "Now he can begin to harvest the trees in the first plot because they have grown to be quite big. He cuts down all the trees. The trunk of the trees and the big branches are good for making charcoal. You need big thick pieces of wood for that, as thick as your upper leg or even more. The farmer uses the smaller branches to make wood for firewood, about the thickness of the lower leg or arm. He cuts them into pieces about a meter (3 feet) long and ties them in bundles. He puts the bundles in a shed until they are dry and can bum well. After cutting down the trees, he burns that first hectare plot (2.5 acres), being careful not to let the fire get into the other plots where the trees are growing."

Joel asked, "Why does he bum the soil? Isn't that bad for it?"

"No," the Professor replied. "Many acacia seeds have fallen into the ground from the growing trees. The fire burns open the hard shell on them. That allows thc seed inside to germinate and sprout. So burning the field is a good thing."

"Does he then get new seed from the agricultural school?" Amos asked.

"No. Now he doesn't need more seed, because lots of seeds are already in the ground."

"But how does he get them to grow in rows?" Obadiah asked.

"That's easy," the Professor said. "John, tell them about that."

John said. "The seeds are scattered everywhere and they soon sprout up. Many young trees sprout up, far more than is necessary. So the farmer pulls out many new trees, leaving others to grow that are about three meters (10 feet) apart. If trees grow very close together, they will not grow well."

Amos said, "So you are saying that growing acacia trees and planting crops can go along together?"

"Absolutely," the Professor replied. "It is really quite remarkable. The acacia trees put lots of nitrogen into the soil. This makes the crops grow very well. The production of maize, millet, cassava, beans, etc. is much better because of this. The farmer plants his crops in the fields where the trees have just been cut down and after the ground has been burned. The crops can grow there between the new trees until they are two years old."

"Can the acacia branches be used for mulch?"

"Oh yes," the Professor replied. "The ends of the low young branches can be broken off the middle-sized trees and used for mulch between the crops. They must be cut before they grow long thorns. This increases food production. It provides compost for the crops. It also keeps the moisture from the rain in the soil."

Marketing charcoal and firewood

"How do the farmers sell the firewood and charcoal?" Henry asked. "They are very busy, so how do they have time to get it to the market?"

"Avery good question," the Professor replied. "These farmers work together. They have formed a cooperative. One of the men is the manager of this big project, and there are about forty farmers all together. Each farmer brings to the manager the sacks of charcoal and the bundles of firewood he has made. His name is on each sack and bundle. The manager makes a note of them all – to whom they belong and how many there are. The merchants from the city come to the manager. He now sells them as many sacks or bundles as they can buy. He makes note of whose sacks and bundles he sells and the money that he receives for them. The farmer then can come and get the money from the sale of his firewood and sacks of charcoal."

Stacks of wood for firewood and larger logs to make charcoal.

"But how does the manager make a living?" Amos asked.

"The manager keeps three percent of the sale price of each sack and bundle. He does not grow trees of his own. Since he gets three percent of the sale of the sacks and bundles from each of the forty farmers, he makes as much as they do."

"Where does he grow his food?"

"The farmers near his home let him use one or more hectares of their land where they have recently cut down trees. So the crops of the manager also grow well in the rich soil where the acacia trees have grown."

Amos asked, "How do the farmers know the manager is giving them all their money?"

The Professor laughed and said, "You have asked a good question. Each farmer keeps his own record of what he has given to the manager. He gets a receipt for the wood he has brought. He knows how much money he should get. All the farmers and the manager meet once a month to discuss the project. They look at all the records. Each one can see what he has done, and how much money he gets and should get. That helps keep everyone honest."

A market community

"They have another project as well. John, tell them about that."

John stood up and spoke to the group. "The flowers of the acacia tree make good honey. Several of the farmers' wives, or older children, have built bee hives and have learned how to keep them and make honey. It has a wonderful taste. They sell it to the merchants who come from the city. It is another way to make money."

"What other things have they learned from this?" Joel asked. "Small animals now live in the acacia forest, and they provide some food. Mushrooms grow very well, so the people have lots of mushrooms to eat. Even a few hardwood trees are growing, and they let them grow. They will provide money for their children in the years to come."

"This sounds like a wonderful arrangement," several of the men said.

"It is," the Professor replied. "One of the merchants has built a store in this project. The people can now buy things right there instead of going into the city for them. They have also built a small health center. A nurse lives there, so they get the health care they need. They have built a small church. One of the farmers is the pastor and he has a smaller tree farm. That way he makes much of his money from his firewood and charcoal and the crops he sells. But he also has time to be the pastor. They have likewise built a primary school for their children."

Amos, the headman of Kitaka, asked a question. "We don't have enough land for a big project like that. Some people live in houses in the community. Others have small farms around the town. How can we get firewood for the people in the town and the farmers themselves?"

Growing trees for fuel in the community

Pastor Simon stood up and asked permission to answer that question. The Commissioner nodded, and the pastor began to speak. "This is a very practical question for most of us. It is especially important for church leaders to understand, for we can help the people in our communities with this.

"I live in the town of Katindi, and I have a small farm just outside. I have planted leucaena and acacia trees around my farm, and acacia trees grow near my house. The trees around my farm make a living fence. The roots of those trees make the soil on my farm rich. I break off many small branches regularly and use them for compost.

I also cut off branches from the taller trees, branches about the thickness of my arm. I use them for firewood.

"I have a little place right in back of my house where I keep those branches. I built a little shed and covered it with a sheet of plastic and palm branches on top to keep out the rain. I put the branches there to dry out for a few months until they are dry enough for firewood.

"Every farmer can do this. It is not necessary to grow eucalyptus or black wattle trees for firewood in our communities. Even people in towns who do not have farms can have two or three acacia or leucaena trees to produce firewood. They can even use branches of mango or other fruit trees to dry for firewood."

Henry spoke up. "Thank you, Pastor Simon, for that good word. A few farmers around my town of Kimbata have part of their land in trees for firewood. Some are growing acacia, like the people at Mampo. Others are growing leucaena, grevillea, or other kinds of trees. They make good money selling firewood to the people in town."

Pastor Simon asked for permission to speak again. "Mr. Commissioner and Professor Kingwedi, here are marvelous examples of real community. The people are providing something that others need – charcoal, firewood, and even food. They make money from what they produce. The people at Mampo have organized a good system that works well. They even have a church, a health center, and a primary school. The people in Kimbata and in our town of Katindi are growing more food to eat and sell to others. We are also growing trees for fuel. We are trying to do what God has told us to do:

- take charge of our own lives
- work together to help one another
- live together in peace and harmony."

Joel asked the Professor, "Sir, would it be possible for some of us leaders to go visit that project at Mampo? It sounds like we could learn something from them."

Starting a fuel tree project

The Professor answered, "Yes indeed. If you think this is an example you would like to follow, we can organize such a trip if you wish. It is quite some distance from here, so we will need to plan it a few weeks ahead. I will let you all know about it.

"However, in the meantime, I would like to invite any of you to come to our agricultural school to get advice about growing trees. A teacher could visit each of your communities. He would need to test the soil. He could then determine which species of tree would be best for you. Acacia auriculiformis grows well at Mampo. That is the species our teachers told them to grow. But it might not grow well in your soil. It may require another species of acacia, or leucaena, or another type of tree adapted to your soil. So let us help you determine which tree and which species is best."

The Commissioner now stood up to speak. He thanked the Professor for his excellent teaching. He thanked everyone for coming and participating in the discussion. He also thanked Pastor Simon and the other church leaders for being there. He said there are one or more churches in every community. These churches should work together to teach people about the land, the trees, and the forest.

"If you work together, and work with the experts from the agricultural school and with us from the government, we can truly restore our land and forests and have lives that are prosperous and happy."

Questions for study

1. *How can people produce wood for fuel without destroying the forest?*

2. *Explain how trees for fuel and crops can grow well together.*

3. *Is it possible to start a project like that at Mampo, even with a small group of farmers?*

4. *What does it require to make a project like this function well for many years?*

5. *How can people in a town or village grow trees for firewood?*

6. *What can you do in your community to get people to work together to do this?*

Chapter 19

Land, Water, and Forests

A week after the meeting in the government center about growing trees for fuel, Moses walked over to Katindi. He had learned many things already from Pastor Simon. He saw what the people in Katindi were doing to grow more food. He felt the people of his village could do these same things. He had another question he wanted to ask Pastor Simon, a question about water.

Having plenty of clean and water

"Pastor Simon. Do the people in your community have enough water to drink? And is it clean water?"

"Yes," replied Pastor Simon. "We have two good springs that give us clear water. We have protected those springs carefully so that the water remains clean."

"How do you protect a spring?" Moses asked.

"First of all, we built a strong dam of rocks and soil about 3 feet, or one meter high around each spring. Clear spring water fills up the little pool or reservoir behind the dam. We put short lengths of bamboo through the top of the dam. The clear water from the top of the reservoir flows through these pipes.

"Next, we built a strong fence all around each spring. The fence goes across the top of the dam. This is to keep people and animals out of the reservoir. Twice a year a man goes into the reservoir to clear out the leaves and other stuff that have fallen into the water. This is to keep the reservoir clean."

"Are you sure the water coming through those bamboo pipes is clean?"

"Absolutely," the pastor replied. "The water from the spring comes from deep down in the hillside. We know it is free from contamination. It has come through many feet of soil and sand that filter out everything harmful. The strong fence keeps people and animals out of the reservoir so the water does not become dirty. Now tell me why you are asking about water."

A protected spring of water.

Moses answered, "Because we do not have enough water in our village. We had a good spring that gave us lots of water many years ago. But for several years the spring is giving us only a little water. We do not know why it is giving us so little water. We also don't know what to do about it."

"Moses," said Pastor Simon. "Come to my house for dinner this evening. We will talk about this then."

A village without water

After eating dinner, the pastor, Virginia, and Moses sat outside around a small fire and talked. Robert joined them. Pastor Simon knew Robert had a story to tell about water, and the pastor wanted Moses to hear it.

Robert said, "My father was a school inspector. Twenty years ago he visited a village called Katunda quite far away. It had no spring of water at all. The women had to walk 4 miles (7 kilometers) to get water. My father asked the chief why they had put the village where there was no spring. The chief replied that their ancestors a few generations ago had built the village there.

"'Was there no spring here when the ancestors established the village?' my father inquired.

"'Yes there was. It was down in the hollow place in the meadow over there,' the chief replied, pointing to the bottom of a grassy valley.

"My father asked, 'What happened to the spring?'

"The chief said, 'Apparently the ancestors were upset with us. The people supposed it was because women had broken a tribal taboo. So the ancestors, in anger, dried up the spring to punish us.'"

Pastor Simon told Robert to stop for a moment. He looked at Moses and said, "The belief in the power of ancestral spirits is very common among traditional people. We must recognize this deep-seated belief. But we must try to help them understand that it is far better to trust in God who created us and our ancestors. His power is greater than the power of all other spirits. The wisdom we need comes from his Spirit and his Word. We don't need to be afraid of ancestral spirits or any other spirits. And we need to get our wisdom from God."

"So what did your father say to the chief of Katunda?" Moses asked Robert.

Robert replied, "He asked him if their ancestors had left them a forest. He had seen no forest near there.

"'Oh yes,' the chief replied. 'They left us quite a big forest.' He pointed to where the forest used to be, down the hillside in the meadow where the spring had once been. 'But we cut down the forest to make gardens, so we could grow more food,' he said.

"My father said to him, 'I am sure your ancestors were wise. They knew the forest was valuable. They knew that somehow the forest was necessary to keep water in the spring. They left it in your care. But you cut it all down.'

"The chief admitted that was true. But he explained they needed the land on the hillside and down in the valley to grow more food. So they had cut down all the forest.

"My father told the chief that God's Word says where there are no trees, there is no rain and no water. He then suggested they try an experiment; let the forest regrow on the hillside and throughout that whole meadow. They would then see if the spring would come back."

"Why did he suggest that to them?" Moses asked.

"Because there is a close relationship between the forest and springs of water," Robert replied. "Let me explain this to you. I teach the students about this in our school."

The relationship between the forest and springs of water

"We often have heavy rains. When the rain falls on the trees in the forest, it does not fall directly on the ground. It hits the leaves and some of it runs down the branches to the trunk of the tree. This water goes down the trunk and into the ground. It follows the roots of the tree deep down into the ground.

"Much of the rest of the rainwater falls gently from the branches onto the ground. It also goes down into the ground, and follows the roots down deep. The tree roots take some of this water back up into the tree so the tree can grow. The rest of this water infiltrates down into the ground until it comes to the rocks deep down in the ground. It cannot go any deeper because it cannot pass through the rock. It now flows down the surface of the underground rock. It makes its way down to the bottom of the hill where it comes out in the spring."

"You mean the water that comes out of the spring comes from the rain?" Moses asked.

"That's right," Robert said. "It is the trees and the forest that help the water from the rain go deep down into the ground. The water then flows down on the deep rock to the spring. We call this deep water the ***water table***. It is that water that keeps the spring flowing."

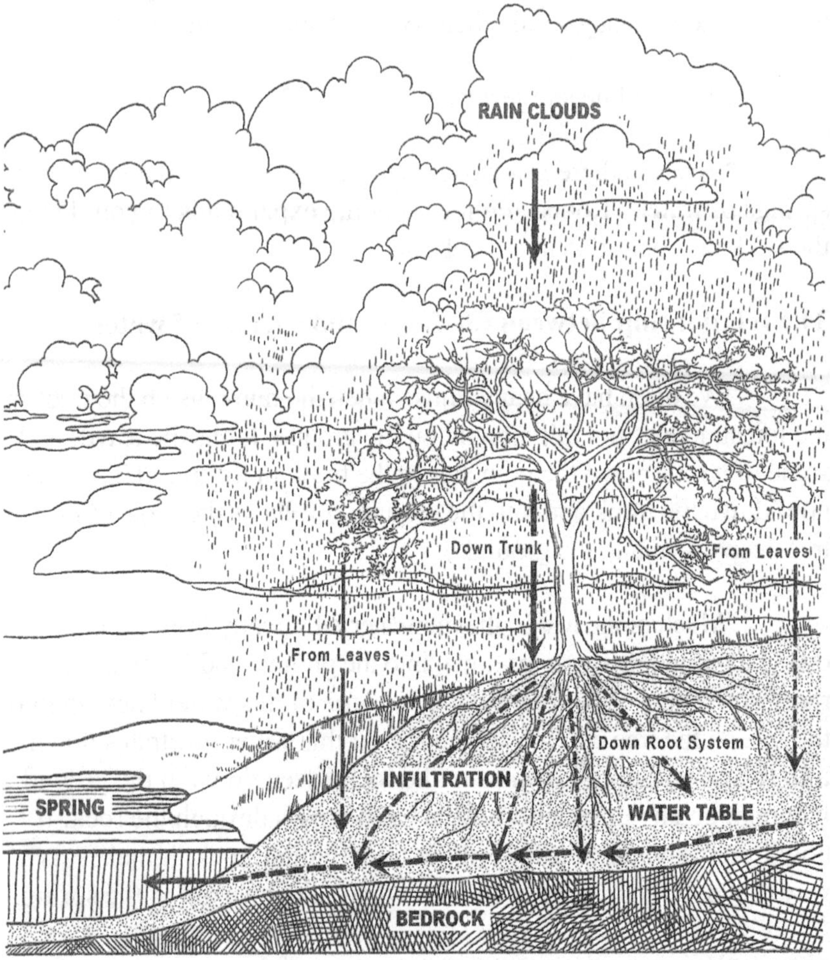

How trees help rain replenish water in the spring.

"Moses, this is very important for you to understand," Pastor Simon said. "If we cut down the forest, the rain will fall on the ground and hit the ground hard. There is nothing on the ground to hold that water. So it quickly runs on top of the ground down into the valley and it is gone. Very little infiltrates into the ground to get into the deep water that comes out in the spring."

"You mean, if we have no forest, there will be little water in the spring?" Moses asked.

"Correct," said Pastor Simon. "Without the forest, the rainwater has no way to infiltrate into the ground and fill up the deep water table. The spring may then dry up altogether."

"That may be our problem," said Moses. "We have very little forest left. Many years ago we had a big forest. But we cut most of it down. We needed to make fields to grow food. We cut down trees for fuel. And we cut down more to build our houses. Now we have almost no forest left for us or our children."

"Where was your forest?" asked Pastor Simon.

"It was on the hillside going down into the valley where our spring is. Now that hillside is almost bare," Moses replied.

"Then you need to do what the people in the village of Katunda did," Pastor Simon said.

Katunda now has abundant water

"Did the people of Katunda really accept your father's suggestion?" Moses asked.

"Oh yes," Robert said. "He went back to their village a few years later. He saw they had a nice forest growing in that valley. To their great pleasure, the spring had returned. They were getting abundant clear water from it."

"Amazing," said Moses. "I wonder if the ancestors knew about that."

"Our ancestors probably did not know about water tables and how rainwater gets down through the ground. But they certainly knew the importance of the forest. I presume they knew the forest had something to do with springs of water."

Moses now asked Pastor Simon, "Could you come to my village next week? I would like you to explain to the people what you have just told me. We can then discuss with them how we can let our forest come back."

Restoring the spring of water at Malenga

Pastor Simon agreed. The next week he went to Malenga to meet with the people. Moses had asked the elders of the village and the leaders of the women to sit together to talk about their problem of having little water. Pastor Simon told them the story of Robert's father about the village of Katunda. He explained how the spring there had dried up after the people cut down all their forest. But now they have allowed the forest to grow back. Once again water was coming from their spring of water.

Moses then spoke to the group and said they must do the same thing. They must make the following rules for everyone to obey:

1. No one could bum the hillside above the spring or the fields in the valley.
2. No one could plant crops anywhere on that hillside or around the spring.
3. When little trees began to grow, no one must cut any of them.
4. When the trees grew tall, no one could cut branches of those trees.

5. This whole area must be left entirely alone until the trees become big and the real forest comes back.

However, some of the elders were not happy about this. One said that, if they allow their forest to come back, people from other villages nearby would come in and cut down the trees. Others disagreed "We all know which land belongs to which village. We still respect that. No one will come in and disturb our forest."

Another elder spoke about a different problem. Each part of the hillside belongs to a different family. Each family needs that land to grow crops because the other fields have become weak. Those families will not agree to let the forest grow back.

Moses explained how the land around the town of Katindi had become fertile once again. He had been there and he had seen what Pastor Simon and the people of Katindi had done. He told how they had planted acacia and leucaena trees that made the soil fertile. He explained crop rotation. He told the elders about Zai holes. He said if they would do all these things, they could use the land on the top of the hill and the surrounding fields. It would not be necessary to use the hillside above the spring.

Pastor Simon then read from the book of Isaiah 41: 17-19.

The poor and needy search for water, but there is none; their tongues are parched with thirst. But I the Lord will answer them; I, the God of Israel, will not forsake them. I will make rivers flow on barren heights, and springs within the valleys ...I will put in the desert the cedar and the acacia, the myrtle and the olive. I will set pines in the wasteland, the fir and the cypress together.

He told how God had commanded the man and the woman to rule over the land. They could use the land to grow crops for food. But they must take care of it and protect the land and the forests. He encouraged them to do what Moses had just told them to do. He assured them God would bless their land, and water would be plentiful in their spring.

The leaders of Malenga spent some time talking about this. They agreed to the rules Chief Moses had announced to them. They said they would inform all of the people. They prayed together to ask God for wisdom to teach the people. They also asked for courage to apply these rules. Moses and the leaders asked God to give them much water from the spring.

Questions for study

1. *Study Isaiah 41: 17-19.*
   - *Why did the poor and needy have no water?*
   - *What had to happen to make rivers flow and streams of water appear?*
   - *Who makes trees to grow?*
   - *Who must do the work to help them grow?*

2. *Why are trees and forest important for springs of water?*

3. *Where does spring water come from?*

4. *How does rainwater get deep down into the ground and then to the spring?*

5. *What happens when there is no forest?*

6. *What are the reasons we have cut down so much forest?*

7. *What must a community do to allow the forest to come back in a certain area?*

8. *Why is it important to obey God's command to protect the land he has given us?*

Chapter 20

Farmers and Cattle Herders

Pastor Simon read the letter carefully. He called Virginia and the two of them discussed it together. The letter was from leaders of a big church group far away. They had local churches in many communities, and many of the members were farmers. However, in the same area was another group of people who had large herds of cattle. These people wandered from one place to another wherever they could find grass for the cattle. Often their cattle would get into the fields and gardens of the farmers and destroy the crops. Much conflict occurred between the farmers and cattle herders. The church leaders had heard about what Pastor Simon was doing with the farmers around Katindi. They were asking him to come to give them advice about what to do to protect their farms from the cattle.

Virginia suggested that he go to visit them to see what the situation was. His nephew John might be able to go with him. She mentioned that John had a close friend, Caleb, who was a teacher in the agricultural school. Caleb taught students about cattle, and perhaps he could go along with them.

Two months later, Pastor Simon, John, and Caleb arrived in the large town of Turgan to meet with leaders of this church group. The leaders warmly welcomed Simon and the other two men and expressed appreciation for their visit.

The next day Pastor Simon, John, and Caleb met with the leaders all day. Several of the local farmers met with them. The farmers told them how they tried to keep the cattle out of their fields. They had built some fences out of sticks and vines. But they had no money to buy expensive fencing material for all of their fields. They had talked with the leaders of the cattle herders group. They insisted they watch their cattle and keep them out of their fields. But the cattle still invaded their fields frequently, and often fights would break out. People sometimes were killed.

Pastor Simon, John, and Caleb visited the farms of several of the farmers around Turgan. They saw that these farmers needed to apply some of the farming practices the people of Katindi were using. The soil seemed weak. No one had planted trees around the fields. The crops looked poor. They saw only a few fences, and they were not strong ones. Pastor Simon suggested to the church leaders that they spend a few days there to teach the farmers what they had learned in Katindi. The leaders were happy to arrange for a farmers' conference for three days. John would stay and do this teaching.

In the meantime, Pastor Simon and Caleb went to visit leaders of the nomadic herders group. One of the local men went with them as an interpreter. This man introduced Simon and Caleb to the leaders. He explained that they had come from a part of the country quite far away. The people there had difficulty finding enough food to eat. But the pastor had discovered ways the people could produce more food. Now their lives were better. He had heard that the people here also had difficulties getting enough food. So he had come to see if he could be of any help to them.

The leaders responded that they raised cattle. They were not farmers. They asked what Pastor Simon knew about cattle. Pastor Simon introduced Caleb to them and said that he knew many things about cattle. Caleb then asked if he could see some of their cattle, and the leaders agreed.

The leaders showed Caleb their herds. Caleb admired their cows and said that many of them looked healthy. He saw many good looking goats. But Caleb also saw many cattle and goats that were thin, and some of them looked sick. But he did not talk about this right away. After spending several hours looking at the herds, the men returned to the community. The leaders agreed to meet with Caleb and Simon the next day to talk about their cattle.

The next morning twenty cattle herders met with Caleb and Simon. They sat together in a circle under a big acacia tree. Caleb congratulated them for the many cows and goats they had. He pointed to a large herd nearby and said many of them looked healthy. He asked, “Do your cows and goats bring you a lot of money?”

One of the cattlemen replied, “We do not kill our cattle often. It takes them a long time to become fat. Many of them never get fat enough to kill for meat.”

Caleb asked, “What problems do you have with your cattle?”

The head man answered. “Our cows often become sick. We do not know how to treat them. Many of them die. Also, we cannot find enough grass for all of our herds. Often we must wander far away to find grass to feed them. The rains do not come any more like they used to do years ago. We don’t know why that is. But we do know that is why the grass does not grow well.”

Another herder spoke up. “We have much trouble with the farmers in this area. They keep trying to take more grazing land from us. They say they need bigger farms. Often we have fights with them because they are taking our pastures.”

Caleb spoke again. "I did see many cows yesterday that looked sick. I can tell the sickness they have. It is quite easy to treat that sickness. In fact I have brought some medicines with me that will help the sick cows get well."

The herders were glad. "Can you treat them now?" they asked.

"We can treat them now," Caleb replied. "But I would like to treat them with you."

"What do you mean?" they asked.

Caleb said, "I would like to teach some of you how to recognize the different sicknesses your cattle get. Then I can show you which medicines to use for those sicknesses. I will teach you how to give the medicine. Since I cannot remain here for a long time, you will be able to treat the cattle yourselves. You can choose a few young men whom I can teach."

"Where can we get the medicines?" they asked.

"There are stores in the district capital that sell medicines for cattle," Caleb replied. We bought some in those stores on our way here. You can buy the medicines there once you know what medicines you need to treat your sick cattle."

Caleb went on. "I saw another disease among your cattle that can be prevented. We have a vaccine that can keep your cows from getting this serious sickness. I have some vaccine with me. I can teach your young men how to vaccinate all of your cattle."

"We are very poor," the head man replied. "We have no money to buy medicines."

Caleb asked Pastor Simon to talk to them about that. They all turned to listen to the pastor.

"In our community of Katindi," Simon began, "our people were also poor. We did not have money to buy the things we needed to grow more food. However, we soon found we could do many things to improve the land that did not cost much money. When we did these things, we began to have more food. We could then sell the extra food. That gave us money we could use to buy other things that would help us grow even more food."

"What can we do here?" the herders asked.

Pastor Simon replied, "You do have some cows that are quite fat. They can produce a lot of meat. When we met with the farmers in Turgan last week they said they could not get much meat to eat. I believe if you would let them know you are going to slaughter a few cows, they would come and buy the meat. You would then have money for the medicines your cattle need."

The head man said, "You have given us good ideas. We will talk them over. This afternoon we will meet again with you to talk more."

That afternoon the group met again with Pastor Simon and Caleb. The head man introduced six young men to Caleb. "These are the young men you can teach how to treat our sick cattle," he said. "You can also teach them how to vaccinate the cattle."

An old man then spoke up. "Pastor Simon, many of the farmers do not like us. If we try to sell meat to them, they will not buy it because they do not trust us. Can you help us with that?"

"Yes, I believe I can," Pastor Simon replied. "We spent a few days with the leaders of the farming people. They trust me. When I explain to them what we have discussed here, they will be happy. They do want to live in peace with you. Caleb can examine carefully the cows you will kill for meat. If he assures them the cows are healthy, they will buy the meat."

This made the herders happy. Caleb said that he would now go to the herds with the young men and begin treating the sick cows.

A week later, Pastor Simon was back in Turgan with the church leaders. John had finished the three-day workshop on good methods of farming. Pastor Simon was pleased to see the people already beginning to do some of the things John had taught them. Some were starting to plant seeds of acacia trees around their fields. Some were making piles of compost to use in their gardens. He saw three families that had made small shelters for nurseries to grow new trees.

The next day three cattle herders came to Turgan and asked for Pastor Simon. They explained to him that they wanted to bring some cows to the town the next week to kill in order to sell the meat. "Have you talked to the people here about that?" they asked.

Pastor Simon took them to the head pastor of the church and also to the mayor of the town. He introduced these herders to them and explained what he had done among the herders. He said that Caleb was still working among them. He would carefully examine the cows they would bring to the town to sell for meat. The mayor insisted that a government inspector look at the cows before they were killed. He would have to approve of their health before they could be killed for meat. Simon, through the interpreter, explained to the herders that this would work well. If the cows were healthy, the people could buy good meat. This would help build trust between the two groups.

Two weeks later Pastor Simon went back to visit the cattle herders. John went with him. The butchering of ten cows in Turgan had gone very well. The farmers were happy to purchase the meat and the herders had returned with much money. Caleb was still with the herders. The young men were quickly learning how to recognize and treat the common cattle diseases. They also had vaccinated three thousand cows.

The head man invited a big group of herders to meet with Pastor Simon and Caleb. Some of them came from quite far away. They all knew the problem of getting enough grassland for pasture. They also knew that even the government was not happy with their constant wandering. Their wandering herds were causing problems with farmers in many places. So they wanted to talk with Pastor Simon and Caleb about this.

Pastor Simon began. "In the beginning God made the world. It was a wonderful world. It was covered with forests and grasslands, with abundant springs of water, lakes, and rivers. Cattle and wild animals were everywhere and had plenty of grass to eat. Then God made man and woman. He told them to rule over the land and the animals. He also told them to take good care of the wonderful things he had created for them. Have you heard this story?"

One of the elders replied, "Yes, we know that God made everything and that it was good. But man did something that made God angry. So he left and went far away. That is why we have great difficulty getting wisdom from God."

Pastor Simon continued. "Yes, I know that man disobeyed God. But God found men to whom he could talk and give his instructions. One of them was Moses, who wrote down what God told him. God told him that if men will obey God's laws and his instructions, they will have a good life. This will help them with cattle and with food. It will also help them live together in peace."

"We would like to know more about that," several of the herders said.

Pastor Simon went on. "God gave intelligence to us, just like God himself is intelligent. He told man to study the land and the animals and to rule over them. That is what we have been doing in Kalindi. This has given us much help. God wants us to plan how to use the forests and the land well. God himself plans, and this world we live in is the result of God and his plan. But, as you say, we have disobeyed God. So we have many difficulties. We do not rule well over the land, the forest, and the animals. We have destroyed much of the forest. As a result, we no longer get much rain. We have allowed the soil to become weak so that it grows only a little food. But now, at Katindi, we are studying the land and the forest and finding out how to restore them. We are also finding out how we all can work together to help the many communities around us."

"So what can we do about finding enough grass for our cattle?" one of the men asked.

"Up until now," Pastor Simon said, "you have let nature rule over you. Where the grass grows, you take your cattle there. When it is gone, you try to find grass somewhere else. You and your cattle use much energy simply walking to find more grass. Now, however, Caleb and I can show you how to have good grass without wandering to find it."

"Tell us about that," they said.

Caleb now spoke up. "You can do two things," he said. "One is to improve your pastures. There are certain grasses you can plant that will grow well, provide good food for your cattle, and make your soil strong. One such grass is passepalum. Another is stylosanthes. At our agricultural school we can help you find these grasses."

"You mean if we plant those grasses, we will not have to wander with our herds looking for grass?" they asked.

"Exactly," Caleb responded. "Now let me tell you a second way. During the rains, grass grows well. Some of this grass can be cut, dried, and stored to feed your cattle during the dry season when there is little grass. That way your cattle can stay where they are. You feed them the dried grass or hay as we call it. The cattle can now grow fat quite fast. They will not be using much energy to walk many miles to find grass. They will have grass where they are."

The head man spoke up. "Caleb, we don't know how to grow grass like that. We are not farmers."

John now said, "I have been working with the farmers around Turgan. They are learning how to make their land better. Soon it will produce more food. They will then not need lots of land for their fields. I also talked with them about growing grass to make hay for cattle. They had never heard about that, but they said they would do it if someone would buy it. So the farmers could grow grass and sell it to you to feed your cattle."

"But we can't buy grass for our cattle!" several of them exclaimed. "Grass is free and it comes from God."

"Something else also comes from God," John responded. "God has given us instructions about how to live together in peace and help one another. Think about this. You just sold a lot of meat to the farmers, and they paid you money. They asked me when you could come back again and sell them more meat. So you have a wonderful way to get money to purchase dried grass, or hay, during the dry season. You will also be able to buy medicines for your cattle. You sell the farmers meat, and they will then sell you hay."

Caleb quickly added, "You can also sell them something else they need. You can sell them the manure from your cattle. The farmers need that to make the land grow more food."

Pastor Simon now smiled and spoke again. "My dear friends, I think we are finding some real wisdom here. God wants all people to live together and help one another. For many many years, you herders have had difficulties with the farmers. You have fought over land. They want the land for fields to grow food. You want it for grass to feed your cattle. Now we see how you can learn to live together, trust each other, and help each other. The farmers grow the hay you need and sell it to you. You sell your meat to them, and also the manure for fertilizer. In this way, you live peacefully with each other and you both will live better."

Two months later a big meeting took place in Turgan. The District Commissioner was there with several of his officials. The mayor of Turgan and leaders from the community and the church were there. The head man and several leaders of the cattle herders had come. The District Commissioner spoke to them all, and a man interpreted his speech for the herders.

"I want to thank you all for coming today. This is an important day in the life of our district. I have received reports from the farmers around Turgan and from the cattle herders to the east that you want to live together in peace. You will each provide things which the others need and will sell those things to them."

The people all nodded their agreement, and the Commissioner continued. "What we must now do is to determine officially which lands will be set aside for you farmers, and which grasslands will be set aside for you herders. So I am asking you each to pick four men who can meet together with me to determine the limits between the grasslands and the farms.

"Before I finish, I do want to say a warm word of thanks to Pastor Simon and his nephew John, and the teacher Caleb who have come from far away to help with this. You three men have shown to our people here how God has instructed us to grow food and raise cattle together. You have shown us how to live in peace and help one another. We are grateful for the wisdom God has given you. We thank you for giving us three months of your time to live among us and show us how to live together."

The following week, in the office of the District Commissioner, leaders of the farmers and the cattle herders sat down together. The met for several days to determine the land for the farmers and for the herders. The District appointed an inspector for the cows to be slaughtered, and a meat inspector to make sure the meat was good. The herders setup a small area near one of their communities to which farmers could come to purchase manure. The herders also began making wagons for their cattle to pull to transport hay from the farmers to the herding communities.

Questions for study

1. *Why were the farmers and herders frequently fighting with each other?*

2. *In what way could the farmers help the herders and make money doing so?*

3. *In what way could the herders help the farmers and make money doing so?*

4. *How would this stop the conflicts and help the people live peacefully together?*

5. *How did God's Word help in making the reconciliation?*

6. *How did the government leaders help in this process?*

Chapter 21

Farming, Churches, and the Bible

Graduation

The time for celebration had come. John was graduating from the agricultural school. For three years he had worked and studied hard. He had done well on his examinations. He was now graduating with distinction. His teachers were impressed with the good field work he did. His reports of his trips to Katindi and the work in the communities there had earned him good grades.

Pastor Simon and Virginia came to the graduation ceremony. They were very proud of what John had accomplished. They celebrated with John and many of his friends. They thanked God for all that John had accomplished.

Return to Katindi

After the graduation celebration, John returned home to Katindi with Pastor Simon and Virginia. Two years had passed since he had been there and worked with his uncle in the communities around Katindi. He was anxious to see what changes had occurred.

John was pleased at what he saw in Katindi. They had finished a wonderful harvest. The granaries were full of millet, maize, and ground nuts. Vegetable gardens were growing well around many houses. The hills around Katindi were green and he could see new trees growing. He saw a new carpenter shop, a man repairing bicycles, and women selling bread. They were all happy and greeted John warmly.

With Pastor Simon he walked around the many fields of the people. Rows of good-looking trees were around each field. The people were keeping them trimmed. It was now dry season and no crops were growing. John could see that the ground was covered with the stalks and vines of the recent crops. This was making good mulch and protecting the ground from the hot sun. He saw that the people understood good farming practices and were following them.

Visit to Malenga

One evening at dinner John asked, "Uncle Simon, what is happening at Malenga?"

The pastor replied, "Tomorrow we can go visit Malenga. I think you will be pleased at what you will see. I will send word to Moses, the headman that we are coming."

The next day John and Pastor Simon walked to Malenga. Moses and several of the community elders greeted them warmly. They congratulated John on his graduation from the agricultural school. "You will make a very good farmer," they said.

They sat down together in a circle under a spreading bougainvillea bush. Some of the women joined them, bringing tea. John asked them how their fields were doing and if they have enough food.

Moses smiled broadly. "John," he replied. 'Thanks to what your uncle has shown us, we have plenty of food. Our fields are doing well, and so are we. We no longer are hungry, and our lives are now happy. When we finish drinking our tea, we will walk around and show you how well we are doing."

After a few minutes Moses led Pastor Simon, John, and some of the community leaders on a tour. John was delighted to see many fields with acacia and leucaena trees growing around the borders. The trees were about six feet (two meters) high, and John could see the people were trimming them regularly. He saw they were laying the trimmed branches on the ground to make mulch. They had made good living fences.

Many of the fields had new Zai holes in them. In some fields, he saw men and women digging new Zai holes.

"Have these Zai holes helped you?" John asked.

"Yes, indeed," one of the men replied. "I have used them now for two years. When we go back, I will take you to my granary. I will show you the big ears of millet I have just harvested. You will see for yourself"

Another man spoke up. "Some of our people still do not believe in Zai holes. But the millet they harvested is quite small. They are now saying they will start digging Zai holes for the next crop."

Moses said, "But soon we will not need Zai holes any longer. We can already see that the border trees are making the soil rich. The mulch is helping as well. Soon our soil will be fertile like the soil in your fields at Katindi."

Nurseries to grow trees

After seeing many good fields in Malenga, they returned to Moses' house. Moses showed them the nursery he had behind his house. They could see many new small tree seedlings growing under the shade of a shelter.

Proudly Moses pointed out: "Here are new fruit trees I am growing. I will plant a few more by those orange and grapefruits trees over there. I will also be able to sell some new fruit trees to my neighbors. Now let's go see our community nursery."

They walked to the primary school. Behind the school was a large shelter with many new small tree seedlings growing under the shade of a shelter.

"This is the second year we have done this," Moses said. "These trees are for planting on the hillside over there. You see those gullies? That is where we have already planted some small forest trees. They are growing well. When the rains return, we will plant these little trees also in those gullies. Soon we will have a new forest filling those gullies and protecting the hill from erosion."

One of the men said he now had two guava trees, three avocado trees, several citrus trees, and bananas and papaya trees around his house. They would soon have much fruit to eat.

John and Pastor Simon were delighted. "You are making wonderful progress," they said. They went back to Moses' house and again sat together under the big bush.

Animals

John said. "I have not seen any goats or pigs wandering around. A few years ago, they were a big problem."

"Yes, indeed, they were," Moses answered. "But now we no longer have that problem. We have several farmers who raise goats and pigs in their farms down in the valley. On your way back to

Katindi, go down into the valley. Stop at Jesse's farm. He will be glad to show you his animal farm."

Moses and some of the leaders went with Pastor Simon and John to the edge of the community. They thanked the pastor and John for their visit. They gave their best wishes to John and told him they expected he will become a great farmer.

Animal farms

The pastor and John took a path down into the valley. Soon they came to the gate of the farmer from Malenga named Jesse. They entered the gate and were impressed by how neat the farm looked. Hedges of cactus were on both sides of the path. They were all around the outside of the farm. They saw Jesse at work down near the stream in the bottom of the valley. Jesse saw them approach and came to greet them. Pastor Simon introduced his nephew John to Jesse. He said that John had just finished his agricultural studies and wanted to see how Jesse raised animals.

Jesse was pleased and invited them to sit in chairs in front of his house. John could see a herd of goats inside a large enclosure of cactus. Jesse explained that he had twenty female goats and three rams. Each year he had many new goats, sometimes as many as one hundred. He kept the little goats in a separate enclosure after they had been weaned. In addition to the grass and forage they found there, he gave them some maize and millet so they would grow fast. He then sold the young male goats and most of the female goats. He made quite a bit of money on the animals he could sell.

John asked him if he got milk from the goat. "No," Jesse replied. 'These are not milk goats. I have thought about getting milk goats. However, I would have to buy equipment to handle the milk. Selling the fresh milk also requires much work. So I decided just to raise goats to sell."

"Do you raise any other animals?" John asked.

"Oh yes," Jesse replied. "Come, I will show you how I raise pigs on my farm down in the valley."

Together they went to the stream. John saw a strong living fence of trees with barbed wire strung between the trees. It went around about an acre (1 /4 hectare) of land. John could see the cut stalks of millet and maize that had just grown there. In the field near the stream was a pen made of wide boards and strong posts. Two big sows were each in a smaller enclosure. Several little pigs from the litter of one sow were in another small enclosure.

The men stood in the shade of a tree to talk. Jesse explained how he brought grain to feed the pigs, most of which he grew in his fields. He also brought down garbage from their house. When the little pigs were grown, he would sell them and make more money.

"Do you have any cows?" John asked.

"No," Jesse answered. "It is difficult to make money from cows. You need almost one hundred cows to make money. That requires up to fifty acres (twenty or more hectares) of land. I have no way of getting that much land. Besides that, I would have to spend much money for medicines for ticks and other diseases. So I have decided to raise just goats and pigs."

As John and the pastor turned to go, Jesse said, "John, it is your Uncle Simon who showed me how to raise goats to make money. Then his friend Roger showed me how to raise pigs. I am very thankful for the help they gave me. I am now doing very well for my family. I have two boys in school, and more of my children will go to school. I think I will have enough money for them all."

The men shook hands. The pastor and John walked back up the hill and returned to Katindi. As they went John thanked his uncle for the wonderful way he has helped many people in Katindi and other communities grow more food, make money, and learn how God wants them to live.

The future

A few days later, in the evening, Pastor Simon, Virginia, and John were sitting in front of the house. The sky was clear. The moon was bright, and many stars were shining. John said, "Uncle Simon and Aunt Virginia, thank you so much for helping me with my studies. You have done much more than just give the money for it. You have also taught me much by your example. I see how dedicated people can make a difference in life.

"I have been doing much thinking since I graduated and came home. I have been praying and asking God what he wants me to do. I have always thought of becoming a farmer, working hard, and making a lot of money. But now I am thinking of something else."

"What is that, John?" Pastor Simon asked.

"I remember when you told us the parable of Jesus about the three servants who each received a sum of money (Matthew 25: 14-30). Two of the men doubled the money, and the master was pleased. The third man simply hid the money and did nothing with it.

"I have learned a lot in agricultural school. I have learned even more from you. You have taught me about farming God's way. And you have shown me how to teach others to do this. I realized that the 'money' in that parable is a symbol of the learning we receive. If we multiply our learning by teaching others, our real Master in heaven will be pleased.

"I've learned from you how important it is to understand God's Word. It tells us how to use the land and how to be a good farmer. I have seen how you, as a pastor, have learned these principles and applied them. I have also seen how you have multiplied your knowledge by helping other people put these principles into practice." He paused and looked up at the sky.

"So what are you thinking of doing?" Virginia asked.

"I am thinking I may go to a pastors' training school like you did, perhaps for a year. I want to learn more from God's Word and about how to teach other people. Then I will go to a community where I can find good land and start a farm. The farm will not be for me to make a lot of money. Instead, it will be a place to teach others how to farm God's way. On the farm I can teach others how to protect the soil and make it fertile. I can show people how to grow different kinds of trees. I may raise some animals to show them how to do that. At the same time, I can teach them what God says about the land, the forest and farming. I can also teach them how to find the life God wants us to have."

Pastor Simon was quiet for a moment. Then he said, "John, you are like a son to us. This would make us very happy. We have tried to help people farm in God's way and to care for the land and resources he has given us. We have helped a few people. I think you will be able to help many more people."

Virginia spoke up. "John, you have learned much about science and agriculture. We did not study those things. We learned simply be observing what was happening and trying to apply the wisdom God gave us. You can now bring together what God says and the Bible and science teach. That can do much to help people know how to live and farm."

Pastor Simon added, "The churches everywhere need this. John, God will use you to help people in churches all over the land to care for the creation, to grow much food in God's way, and to learn the life God wants us to live. You will become a servant to many people."

He opened his Bible to the gospel of John chapter 12 and said, "Here is wonderful counsel Jesus gave us," and he read verses 31 and 32. "*Jesus said to those who believed in him, 'If you obey my teaching, you are really my disciples. You will know the truth, and the truth will set you free.'*

"John, the most important thing is to know and understand what God teaches us and to put that into practice. We then will know God's truth and will be free to live the good life that God intends for us to live."

Pastor Simon and Virginia took John's hands, and the pastor prayed. "Father, we thank you for the life you have given us. We thank you for the land, the soil, and the forests you have given us. We thank you also for how well you have taught John in his school. We ask you to use him to help people all over the country to restore the land and make it produce much food. Give him wisdom how to teach people to live the life God wants them to live. Make him a good servant just like your Son Jesus was when he came to serve us."

The restored land providing abundant food and a foundation for the future.

Questions for study

1. *How does obeying God's Word and knowing God's truth really set us free?*

2. *How do we learn God's truth about farming?*

3. *What does it mean to farm God's way?'*

4. *How did Pastor Simon demonstrate this?*

5. *How can we put together what we learn about the science of farming and what the Bible says about the land and farming?*

Made in the USA
Monee, IL
05 November 2026

68020861R00146